Testing the Data Warehouse
PRACTICUM

Assuring Data Content, Data Structures and Quality

Doug Vucevic & Wayne Yaddow

Order this book online at www.trafford.com
or email orders@trafford.com

Most Trafford titles are also available at major online book retailers.

Printed in the United States of America.

ISBN: 978-1-4669-4356-8 (sc)
ISBN: 978-1-4669-4355-1 (e)

Library of Congress Control Number: 2012914332

Trafford rev. 08/15/2012

 www.trafford.com

North America & international
toll-free: 1 888 232 4444 (USA & Canada)
phone: 250 383 6864 ♦ fax: 812 355 4082

I dedicate this book to the people who are my never-ending inspiration for so many things: to my wife, Ksenija, and our children, Drasko, Diana, and Vukan, and the memory of my parents.

Doug

This book is dedicated to my wife, coworkers, and colleagues from earlier assignments who helped make these writings possible. Included among colleagues are Vincent Rainardi, author of *Building a Data Warehouse with Examples in SQL Server*; Jaiteg Singh, author of *An Introduction to Data Warehouse Testing*; and Raj Kamal (Microsoft Corporation), writer of several papers on data warehouse and BI testing. Thanks so much.

Wayne

Contents

Acknowledgment

First and foremost, we, the authors, are grateful to Quest Software for providing tools and permission to use the screen prints taken with Quest®, TOAD®, and Quest® Code Tester for Oracle. Our gratitude also goes to HP Company for permission to use the screen prints taken with their tools HP Quality Center® and HP Quick Test Professional (QTP)®.

We have been blessed with the great fortune of working at some of the most excellent Canadian and U.S. companies such as IBM; Canadian banks such as Bank of Montreal (BMO), Toronto Dominion (TD), Canadian Imperial Bank of Commerce (CIBC), and Royal Bank (RBC); government of Ontario; Canadian retailers Hudson's Bay Company and Canadian Tire; Canadian aircraft manufacturer Canadair; Standard and Poor's; JP Morgan Chase; and Oppenheimer Funds.

We are most thankful to these great organizations for without the knowledge and the experience that we have acquired at these organizations, the book in this form would have not be possible.

In particular, we are grateful, for sharing their thoughts and suggestions, to the exceptional QA managers Scott Coolling of BMO, David Wu of CIBC, and Wilson MacArthur (ex-IBM manager); Debbie Francis at Oppenheimer Funds; and Steve Labrecque of Canadian Tire Corporation.

Overview

In this new world, information is king. The more information you have, and the better and faster your analysis, the greater the probability that you will make winning investments.

—Geoffrey More, *Living on the Fault Line*

Testing the Data Warehouse is a practical guide for testing and assuring data warehouse (DWH) integrity. It first appeared in the form of handouts that we gave to our students for a course we teach at the Institute for Software Engineering®. It grew out of our frustration while trying in vain to find the appropriate reference material for the data warehouse testing course. We marshaled our own resources, and you are reading the result of it. The book is not based on rigorous scientific evidence. Rather, it is a tale from the trenches of testing battlefields, a message passed from warrior to warrior.

A data warehouse is a valuable corporate asset used to envisage business strategies and make informed business decisions. The enhanced access to information that a data warehouse provides enables an organization to make time-critical business decisions that are required to remain competitive. Data warehousing needs a comprehensive assessment of the impact to the entire organization and development of a plan for an organized, systematic solution.

As for the quality assurance (QA) teams, it creates an exciting new skill opportunity that comes once around infrequently. It is nothing less than a new business paradigm which creates an unlimited learning opportunity (essential if one wishes to prosper in it).

As with any new paradigm, most of us are unprepared for it. That is bad news. The good news—so is everyone else.

The race is on!

The most nimble of us will flourish the most.

Read on!

This book will reward you with a head start.

The enterprise data warehouse (EDW) is a mission-critical asset because it feeds important business intelligence applications used in making strategic business decisions, such as business performance optimization, revenue enrichment, customer service, etc. Defects in the EDW not only increases the cost associated with rework, but also results in lost business opportunities that cannot be known, thus cannot be accounted for or recouped. In view of this, we strive to walk the reader through the testing and quality assurance activities required to minimize the risk of production problems caused by the erroneous use of data. If we are successful with this book, your goal of delivering near problem-free DWH applications will be achieved more easily.

Business knowledge, acquired from EDW, is a result of transforming data into information and finally into business intelligence (BI). The goal of this book is to show an actionable QA methodology and practical testing techniques for delivering near problem-free DWH applications to our organizations. Ours (QA) is the responsibility of ensuring that this technology helps our organizations in maximizing business opportunities by helping them make better decisions and ultimately giving our customers a more rewarding experience. QA professionals must always keep in perspective that DWH application is a solution to a business problem, and if the business problem is not solved for whatever reason—be it incorrect business requirements, wrong design, or coding errors—then the product does not deliver the business benefits it is designed to deliver.

Ours is an era of the global marketplace, and the new differentiator in that marketplace is the effective deployment of decision support technology. The EDW is an enabling intelligence-driven technology. An effectively implemented DWH application can provide a full picture of a business and give insight into the future risks it faces. We are at the historic junction of horse-versus-locomotive competition. Those who capitalize on this new opportunity will emerge as future market leaders.

It is said that the "data is the new oil," but data alone is not enough; it is our ability to create the business knowledge based on that data.

Yes, we are delivering a message of warning, except that our message is not accompanied with despair, but with the hope of a brighter future for all humanity at this critical junction with the new paradigm. We are framing this message into the larger context and relating it to a journey through unknown lands and the stormy seas.

Introduction

The goal of this book is to help the readers effectively plan and conduct the testing of data warehouses, from the profiling of data that is input (source to the data warehouse) to the staging, cleaning, and application of business rules and transformation of specific data elements. In addition, we provide guidance on testing business intelligence reports that use the tested data warehouse.

As the complexities of data warehouse development have evolved, demands placed on database designers, database administrators, and quality professionals have grown and taken on greater relevance. QA teams are expected to check whether data performs in accordance with the intended design and to uncover potential and real problems that were not anticipated in the design. We are, therefore, expected to plan, develop, and execute more tests and be prepared to rerun them multiple times to avoid regression.

In addition, the QA team is expected to provide continuous assessments on the current state of data warehouse projects while under development and after deployment to production.

In order to attain a certain degree of confidence in the quality of the data in the data warehouse, it is necessary to perform a series of tests. There are many components and aspects of the data warehouse that can be tested, and in this book, we focus on the end-to-end ETL procedures.

Due to the complexity of ETL processes, ETL procedure tests are usually custom written, often with a low level of reusability. We address this issue and work toward establishing a generic procedure for integration testing of certain aspects of data warehouse load procedures. In this approach, ETL procedures are treated as a black box and are tested by comparing their input and output datasets. Datasets from three locations are compared—datasets from the relational source(s), datasets from the staging area, and datasets from the data warehouse.

This book is not a pure technical book; rather, it is a technical book framed in a larger context. On the higher level, the book is comprised of three main themes that span across all chapters:

1. Motivation for writing this book.
2. Description and causes of the predicament we are in and its impact on society and each one of us.
3. Exploring solutions to the problem and the opportunity it is creating.

An introductory chapter on the DWH concepts and its components provides a basic explanation of the software you are about to start testing. Good references are provided to the QA professionals interested in pursuing career in this (DWH) fast-growing field of information technology (IT). For a better understanding of data warehousing, we strongly recommend excellent introductory books by Ralph Kimball and Margy Ross, *The Data Warehouse Toolkit: The Complete Guide to Dimensional Modeling* [Ref. 1 end of the book], and *Building the Data Warehouse*, by W. H. Inmon [Ref. 4].

Another rationale for the introductory chapter to the DWH is to show that this knowledge is a prerequisite (as it is the case in testing any software) for the quality assurance (QA) teams that intend to pursue DWH application, QA, and testing.

We offer a brief introduction of business intelligence as this is a primary reason for testing the data warehouse. Deconstructing DWH major obstacles as well as remedies are discussed in the section "Missing Link in BI Success."

After an introduction to data warehousing technology, QA processes and methodology are generally discussed. Differences between software applications testing and DWH testing are considered. Specific strategies for testing DWH applications are recommended. This chapter describes a methodology to deliver near problem-free DWH applications into production. Here, we included a discussion on the data warehousing application testing cycle and how it relates to software development life cycle (SDLC).

New chapters have been added to address planning for data warehouse testing. Among other things, this means developing testing goals and objectives such as assuring complete loads of data in each stage, assuring that data transformations are correct, verifying data quality from source and into target and much more. One section in particular—QA checklists for data warehouse quality verifications—lists tests and checks that most QA teams will want to consider as they profile data, load to staging, transform incoming data, and load to the data mart.

Even though the majority of the book is devoted to the what and how of data warehouse and ETL testing, other sections are devoted to the *why* test and verify.

The great lives in human history have been built on *why*. If the person knows why, she or he will learn how, despite all the obstacles. The key in achieving most anything is not how, but why. A section on risk and review of QA strategy devote considerable effort discussing motivation and positioning DWH application correctly within an organization.

Since most data warehouses will inevitably need to be regression tested after ETL changes, new data integrations, and data corrections have been applied to data, we added a section to highlight approaches to database regression testing.

Automating data warehouse testing can be a significant challenge and although some portions of DWH are amenable. Whether the warehouse is developed in an agile environment or not, automating certain portions of the testing process can pay big dividends.

The last chapter is where we demonstrate with hands-on examples to illustrate developing and executing test cases with the focus on various testing techniques that may be employed in testing EDW. The case study for concept was developed and demonstrated with an example of using automation test tools for regression testing. Software tools like Quest TOAD®, HP Quality Center®, and Quick Test Professional (QTP)® used to illustrate typical real life environment. We also present end-to-end user test form data sources to BI, testing correctness of the reporting and the analysis tool.

Introduction to Data Warehousing Application

Platonic view of the DWH application: Plato focused on the world of ideas that lay beyond those tangibles things. For Plato, the only thing that was lasting being was an *idea*. He believed that the most important things in human existence ware beyond what the eye can see and the hand can touch. Plato debated with his disciple, Aristotle, that influence of ideas transcends the world of tangible things. Plato's reality of whole was greater than sum of its tangible parts. DWH applications, too, the exceed sum of its parts (software, hardware, and data), and its use is limited only by ideas and human imagination. In other words, the DWH application is more than meets the eye.

Data Warehouse

The customer-centered enterprise regards every record of an interaction with a client or prospective client, such as each call to customer support, each point-of-sale transaction, each catalogue order, and each visit to a company web site as a *learning opportunity*. Organizations gather hundreds of terabytes of data from and about their customers without learning anything. Data is gathered just because it is needed for some operational reasons, such as billing or inventory control. Once it has served that purpose, it is left on disk or tape, or is discarded.

For learning to take place, data from many sources such as billing records, scanner data, registration forms, applications, call records, coupon redemptions, and surveys must first be collected and organized in a consistent and useful manner into a system called *data warehousing*.

Data warehousing allows the enterprise to have collective memory of what it has been observed about its customers. Data warehousing collects data from many different sources in a standard format with consistent field definitions, with a single purpose of supporting decision support process.

What is a data warehouse?

A data warehouse is a database in which collected and consolidated data are periodically stored from the source systems into a dimensional or normalized data store. It usually keeps years of history and is queried for business intelligence or other analytical activities. Data is typically updated in batches, not in real time, as transactions happen in the source system.

Sins of the past times:

Data warehouse (DWH) is a relatively recent (1990) phenomenon. Data silos were used (and still are) preceding the appearance of DWH. Each business unit developed its own strategy for managing and using data. Within the same organization, some used Microsoft® SQL Server®, some preferred Oracle®, yet others managed their database on mainframe in DB2. Each business unit has its one definition of *clean data*. But on the corporate level, data from overseas sales, for example, looked quite different from North American sales data.

In the absence of standards across the organization each division entered data in self-serving interest, entering only what is important to them. Different platforms (Windows®, UNIX and mainframe) with different code pages (ASCII, Unicode and EBCDIC) caused the same character to be represented in a different way in each platform. Organizations hired highly skilled professional analysts, typically MBAs, to sort out the problem cause by diversified data. Analysts had to pull data manually and cross-check them over multiple systems to reconcile data and create reports.

Complexity of this process precluded real-time reporting. The consistency of process was affected if an analyst moves to another position; new analysts had different methods, especially when making judgment calls. Problems became evident on the corporate level. Many analysts across an organization created reports with their own "version of the truth," which may be perfectly correct from that division's point of view, but all these reports, when viewed at the corporate level, appeared inconsistent as if they were presenting "many versions of truths."

Building the DWH for more efficient decisions support system

The hunger for integrated corporate data cannot be satisfied within data silos paradigm as the corporate information cannot be easily obtained by adding together the information from many tiny little applications. Instead data has to be recast into the integrated corporate collection of information, called enterprise data warehouse (EDW). The data warehouse represented a major change in thinking for the IT professional.

A DWH consists of a collection of data with purpose of supporting the management's decision. Data from various source systems are collected and consolidated into a usually dimensional or normalized data store to be analyzed. Enterprise data warehouse is an enterprise information environment, a new paradigm with the specific intention of providing vital strategic information.

Most of a company's data is collected in order to handle the company's ongoing business. This is called *operational data* and includes categories such as CRM (customer relationship management) systems, SCM (supply chain management) and databases containing various transactions. The system from which data is collected contains the operational data; hence, it is referred to as OLTP (online transaction processing). A retail database with information about customers, transactions, products, and prices is an example of the operational database. An operational database is used to well-defined questions, such as *what is the total price of the basket*.

A DWH is a system of data that integrates an organization's historical and heterogeneous data into an information source which enables online analytical processing (OLAP).

The OLAP supports different types of queries, aggregation being the most important. A typical query may be: *What are the sales by product, by region, this month in comparison with the same month previous year?*

A data warehouse (DWH) is a database system in which data is collected to be analyzed. An enterprise data warehouse (EDW) is an information environment, a new paradigm with the specific intention of providing vital strategic information.

Most of a company's data is collected in order to handle the company's ongoing business. This is called *operational data* and includes categories such as CRM (customer relationship management) systems, SCM (supply chain management), and databases containing various transactions. The system from which data is collected contains the operational data; hence it is referred to as OLTP (online transaction processing).

The decision support system (DSS or DWH) provides a good physical separation from its OLTP. The DWH is a tool that integrates an organization's historical and heterogeneous data into an information source which enables online analytical processing (OLAP).

The term *data warehouse* actually refers to a collection of relevant data from multiple sources that is rationalized, summarized, and catalogued in stable, long-term data storage, facilitating the management's decision-making process. The major characteristics of DWH are the following:

- *Subject-oriented*—data that provides information about a particular subject, instead of a company's ongoing operations.
- *Time-variant*—all data in a DWH is identified with a particular time period.
- *Integrated*—data is gathered from various sources and merged into a coherent whole.
- *Nonvolatile*—data is never destroyed.

The subject-oriented data of a DWH is organized around the functions of the organization. Information in a DWH is organized into various *dimensions*. For example, for the retail company in our case study at the end of this book, major subject areas—dimensions—might be products, orders, vendors, sales, customers, etc. A sales analysis database is organized according to products, time, territory, and other dimensions. An invoice database could use time, customer, product, and supplier dimensions. Each type of company has its own unique set of subjects.

In practical terms, a data warehouse is a collection of technologies that enable business users[1], such as financial experts, planners, executives and various analysts, to make faster and better strategic decisions. Data warehousing products encompass numerous suites of products for analysis, query, and reporting capabilities that also include DSS characteristics. They consist of software applications that maintain the database and present the data. Metadata—data that describes the data—is usually is stored in a Metadata Repository, but it is also part of every DWH.

The purpose of the EDW is to provide the business value of the data to organizations. The business value of the EDW is directly relevant to data accuracy and information quality, which will be addressed later in this book.

DWH is a denormalized, business-centric *star schema*—a standard data warehouse design with one or more *fact* tables comprising the hub of each star, surrounded by various *dimension* tables that allow the level of granularity of the *facts* to be drilled into or rolled up along relevant vectors. When we are talking about DWH we are most commonly referring to the Star-Join Schema, which consists of:

- *Fact Tables*—What are we measuring?
- *Dimension Tables*—What are we measuring with?

---

[1] The person using DW application; the person interacting with the software

This is quite different from the fully normalized relational architecture adopted by most OLTP systems. The star schema does not take data redundancy into consideration as quick response time is the highest priority. The schema is denormalized; therefore, accessing and filtering huge data sets is extremely fast, since table joins are kept to a minimum and data access paths are fully preplanned in accordance with current business needs.

Robust indexing, which normally has an impact on performance during *insert, delete* or *update* transactions does not affect performance, since only *select* transactions by end-users are allowed. This also eliminates the kinds of *insert, delete* and *update* anomalies inherent to denormalized schemas.

According to computer scientist Bill Inmon's Time-Variant concept, every unit of data in the DWH is accurate at some point in time. In many cases, a record is time-stamped, or it has a transaction date. In every case, it has some kind of time marking to show the time when the record was accurate. It calls for storage of multiple copies of the underlying detail in aggregations of differing periodicity and/or timeframes. DWH may have details for seven years along with weekly, monthly, and quarterly aggregates of differing duration. The time variant strategy is pivotal not only for performance, but also for maintaining the consistency of reported summaries.

Integrated data is the most important aspect of the DWH. Integration means that data from a variety of sources is not comingled or blended together. Integration is the process of mapping dissimilar codes to a common base, developing consistent data element presentations, and delivering this standardized data as broadly as possible. The DWH is fed from multiple and disparate sources of data. Data is then transformed and reformatted into a single corporate image that can be used by decision makers on the corporate level.

Nonvolatility of data literally means that once a row is written, it is never modified. This is necessary to preserve incremental net change history. This, in turn, is required to represent data as of any point in time. When you update a data row, you destroy information. You can never recreate a fact or total that included the unmodified data. Maintaining institutional memory is one of the higher goals of data warehousing. Data in the operational database is regularly accessed and modified one record at a time. Data in the DWH is usually mass-loaded, but it is not changed.

A basic assumption of Inmon is that a data warehouse is exclusively a store of data for decision support. Under his definition, this precludes the use of a data warehouse for what he calls the Operational Reporting Process[2]. In recent years, DWH has

---

[2]   INMON, W.H.: 'Building the data warehouse' John Wiley and Sons, September 1998,41, 9, pp. 52-60 New York, 1997, 2nd Edition

become the main trend in corporate computing, as it provides managers with the most accurate and relevant information to improve strategic decisions.

Planning and analysis applications called online analytical processing (OLAP) are the heart of a data warehouse. The term MOLAP (multi-dimensional OLAP) is also used because unlike OLTP entity-relationship models consisting of two-dimensional tables, data warehouses use a multi-dimensional model for storing data. The term OLAP is neither a meaningful definition nor a description of what an OLAP is.

An OLAP is a series of protocols used mainly for business reporting. With OLAP, businesses can analyze data in all manner of ways, including budgeting, planning, simulation, DWH reporting, and trend analysis.

An OLAP is a type of software technology that gives a company's various managers and executives a capability for quick and interactive analysis of multidimensional information—derived data which has been transformed from raw, primitive data.

Below are some differences between derived and primitive data based on W. Inmon's recommendations:

- Operational data is primitive, whereas DSS data is derived.
- Primitive data is detailed data used to run day-to-day operations. Derived data is summarized for the DSS needs of the company.
- Primitive data can be modified. Derived data can be recalculated but cannot be changed.
- Primitive data is current value data. Derived data is historical data.
- Primitive data is operated on by repetitive procedure. Derived data is operated on heuristic programs and procedure.
- Primitive data supports the clerical function. Derived data supports the management decision making function.

Operational databases have a high number of transactions that take place every hour. This kind of database is always up-to-date and represents a snapshot of the current business state at a given point in time.

The DWH, on the other hand, is an informational database and is static and stable. DWH is refreshed, usually nightly, and represents a snapshot of the business at a given point in time in the past. This is why DWH is not used to make operational, but strategic decisions, which are based on events in the past.

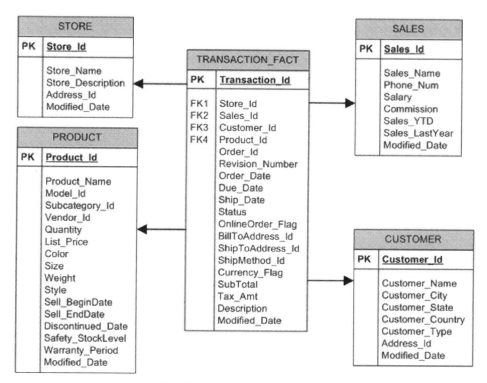

Fig 3.1 Star-Join Schema

The fact table is at the center of the data model. Fact tables contain keys to all the dimension tables and measurable facts required by data analysts. For example, a store selling computer parts might have a fact table recording the sale of each item. Measurable numeric facts (numeric attributes such as sales dollars, invoice amount, etc.) are stored in fact tables. Fact tables can grow very large, with millions or even billions of rows. It is important to identify the lowest level of facts that it makes sense to analyze for the business; this is often referred to as the fact table *grain* which determines the size of the fact table. To access multi-dimensional information, the fact table also contains all *foreign keys*, which are used to join the facts to all *dimension tables*.

Dimension tables organize and index the data that is stored in a fact table. A visual representation of this connection between fact tables and dimension tables appears as a star, hence the name. Dimension tables are typically small, ranging from a few to several thousand rows. However, dimensions can occasionally grow large; a large bank could have a customer dimension with millions of rows. A dimension table structure is typically very shallow; for example, the customer dimension could look like this:

    Customer_ID
    Customer_name
    Customer_city
    Customer_state
    Customer_country

Summarization, or creating rollup summary tables, can be set up with a *materialized view*. These summary tables help avoid heavy queries on the fact table.

Examples of the most common basis for summarization and relation to the fact and dimension tables	
Measures	Allow changes to the view of data, such as Periodic, Week To Date [WTD], Month To Date [MTD], Quarter To Date [QTD], Year To Date [YTD]. Daily transactions are aggregated to provide consolidated weekly or monthly comparisons viewed using a Calendar interface
Dimensions	Services or Products and their properties [COLOR or SIZE] can be totaled into a hierarchy of BRAND, MANUFACTURER, CATEGORY, or other aggregate.
Currency	"Local Currency" and/or "Reporting Currency"

Every dimension, such as time, can be structured into a hierarchy of consolidation levels—years, quarters, months, weeks, individual days, or other levels of data granularity. But *day of the week* is an extended attribute.

The lower (finer, more detailed) the level of granularity available for analysis, the more costly it is to store and process the data.

One other important aspect of EDW is its very large size. For example, Wal-Mart's data warehouse stores 570 terabytes. A terabyte is equivalent to 1024 gigabytes (the prefix *tera* is derived from the Greek word for monster). Google data warehouse stores about 4 petabytes. A petabyte is equivalent to 1024 terabytes (the metric prefix *peta* means 10^{15}).

A DWH is planned and produced to support the decision-making process in an organization. Data are obtained from the production databases and reorganized in the DWH, so that queries can be efficiently answered without hindering the performance and consistency of the production systems. DWH is a tool from where businesses, using applications, obtain the data to provide the information that will contribute to the business's success.

Data warehouses are optimized for speed of data retrieval. The data in data warehouses are denormalized as a consequence of a dimension-based model. This is necessitated to satisfy speed of data retrieval. Data in DWH are stored multiple times—in their most granular form and in summarized forms called aggregates.

	OLTP System (Online Transaction Processing) Operational System	OLAP System (Online Analytical Processing) Data Warehouse
Source of data	Operational data; OLTPs are the original source of the data.	Consolidation data; OLAP data comes from the various OLTP Databases
Purpose of data	Execution of fundamental business tasks	To help with planning, problem solving, and decision support
Application	Operational: ERP, CRM Legacy apps,	Management Information System Decision Support System
Users	Staff	Managers, Executives
What the data	Reveals a snapshot of ongoing business processes	Multi-dimensional views of various kinds of business activities
Inserts and Updates	Short and fast inserts and updates initiated by end users	Periodic long-running batch jobs refresh the data
Queries	Relatively standardized and simple queries Returning relatively few records	Often complex queries involving aggregations
Processing Speed	Typically very fast	Depending on the amount of data involved; batch data refreshes and complex queries may take many hours; query speed can be improved by creating indexes
Emphasis	Insert, Update, Delete	Select-Retrieval
Space Requirements	Can be relatively small if historical data is archived	Larger due to the existence of aggregation structures and history data; requires more indexes than OLTP
Database Design	Highly normalized with many tables	Typically denormalized with fewer tables; use of star and/or snowflake schemas
Backup and Recovery	Backup religiously; operational data is critical to run the business, data loss is likely to entail significant monetary loss and legal liability	Instead of regular backups, some environments may consider simply reloading the OLTP data as a recovery method

ETL—Extract, Transform, Load

ETL stands for extract, transform, and load. It incorporates the dispersed data from all departments within an organization into a uniform DWH. Most often, ETL systems stage the data once or twice between the source and the data warehouse target. In our example at the end of the book, we have one stage, and in this case, the stage is the database (often, the staging area is in the file system).

Extraction is the process of moving operational data into the DWH. Data from online transaction processing (OLTP) and legacy systems provide inflow into the staging servers of a data warehouse.

In order to build the data warehouse, the appropriate data must be located. Typically, this will involve the current OLTP system, where the *day-to-day* information about the business resides, and historical data for prior periods, which may be contained in some form of *legacy* system.

Often, these legacy systems are not relational databases; much effort is required to extract the appropriate data. For example, in a bank, data may be gathered from loan processing, passbook processing, and accounting systems. In a retail store, data may be gathered from point-of-sale devices and cash registers.

Operational data can be in the form of flat files, table records of the relational database, Excel files, etc. The extract program rummages through a filing or database system using specific criteria for selecting or filtering qualified data for extraction and transformation into the other filing or database systems.

In order to find qualified data, many files and formats have to be analyzed. Some of these files are stored in an information management system (IMS); others use the virtual storage access method (VSAM), while some use the integrated data management system (IDSAM). Diverse skills are required for assessing and testing in these environments. An additional complication can be the fact that the element may be stored in two or more of these files under a different name. For example, the entity *Customer* may be stored in another collection of data in which the same entity is known under the name of *Client*. Every piece of data must be analyzed and *rationalized*; otherwise, we will be mixing *apples and oranges* in our reports.

An advantage achieved by the extracting process is that data is copied away from the high-performance operational online processing environment so it does not affect the performance of that environment.

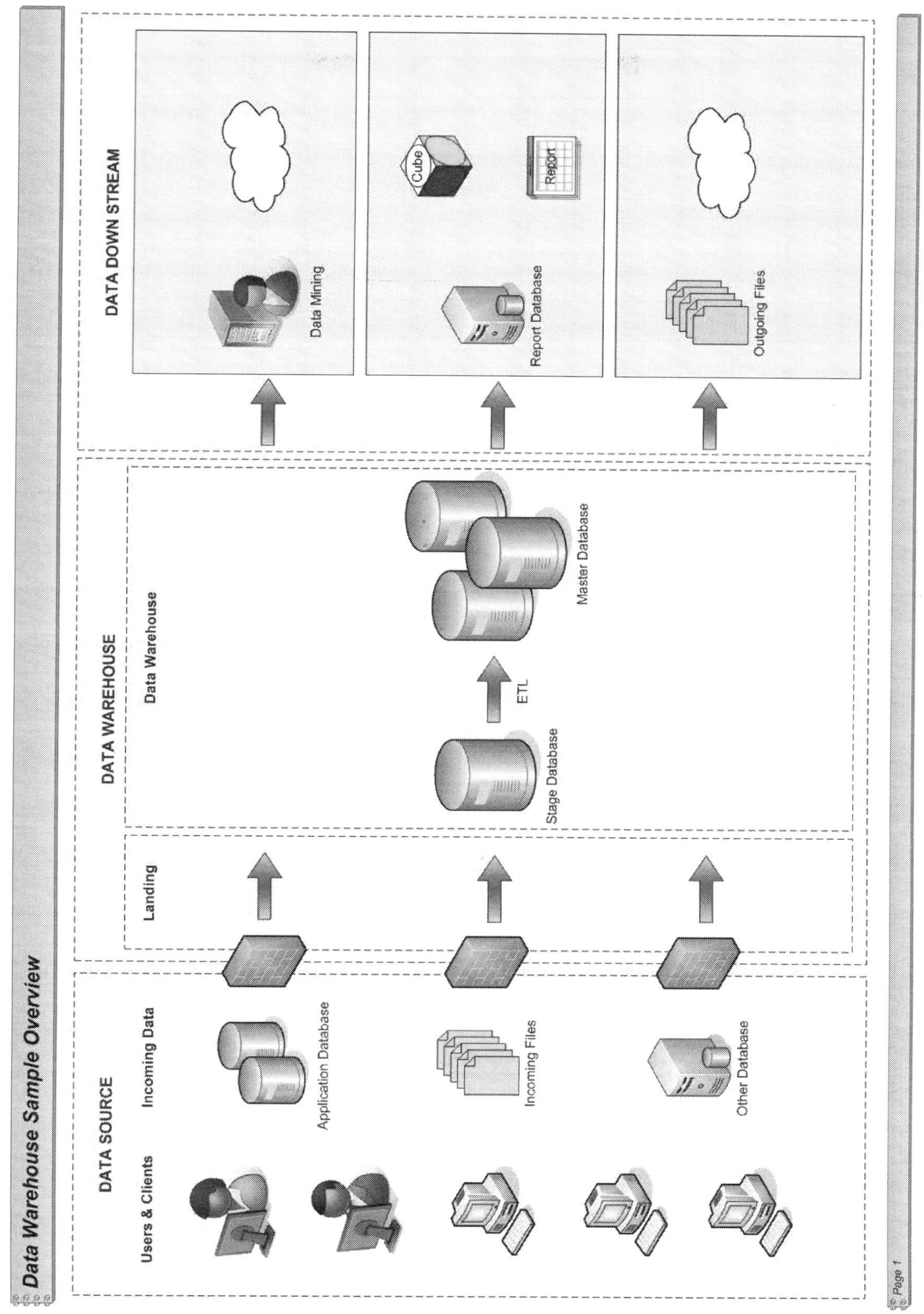

Fig. 3.2 ETL Environment

Transformation is carried out after the DWH schema has been designed. It is a process of changing the data from being transaction-suitable to a structure that is most suitable for decision support analysis.

Data extraction includes a range of data grooming actions which are performed to make data presentable for DWH. Typically, this is performed through a process known as *data cleansing*, usually in conjunction with the data transformation phase of ETL. A data warehouse that contains incorrect data is not only useless, but also very dangerous. The whole idea behind a data warehouse is to enable decision making. If high-level decisions are made based on incorrect data in the warehouse, the company could suffer severe consequences.

Data cleansing is a process that validates and corrects the data before it is inserted into the warehouse. For example, the company could have three *Customer Name* entries in its various source systems, one entered as *RBC*, one as *R.B.C.*, and one as *Royal Bank of Canada*. Obviously, these refer to the same customer. A business decision must be made as to which is correct, and then the data cleansing tool will change the others to match the business rule. The transformation process includes standardizing, integrating, cleansing, augmenting, aggregating, and creating the data sets for loading into the repository. This is why data transformation is the most complex and biggest challenge in building DWH.

Load deals with loading the DWH with the previously transformed data. The DWH has to be populated automatically, usually nightly, and consistently to reflect changes on the operational systems. Loading is a straightforward process. The key consideration in the loading process is to achieve the appropriate speed of data loading. This is achieved by using various methods, such as:

- Presorting the file as per the primary key index for data warehouse loading. This is the most highly recommended technique.
- Turning off Logging during loading.
- Dropping and recreating Indexes. This, as well as turning off Logging, can save on the overhead during the loading.

Conclusion

A relational data warehouse, also known as star schema or dimensional model, is a consolidated, consistent, historical, read-only database storing data from many source systems. The data often comes from OLTP systems, but may also come from files (Excel, CSV, XML . . .)

Data in the DWH is formatted to facilitate the fast response to user's queries. Star schema provides the quick response by renormalizing dimension tables and employing many indexes. Star schema is the best source, as the data is already consolidated and made consistent for cube-building.

Turning Data into Information with the DWH

BI—Business Intelligence

Water, water everywhere, nor any drop to drink . . .

—Samuel Taylor Coleridge

This famous line by Samuel Taylor Coleridge from *The Rime of the Ancient Mariner* is reminiscent of the state of BI in most organizations today. Despite an abundance of data, the staff is often thirsty for information. We are in the midst of an *information explosion* in which there is too much information for anyone to absorb and analyze. The amount of data available has increased dramatically in the last few years, but the ability to make sense of it has increased little, if at all.

> Our networks are awash in data. A little of it is information. A smidgen of this shows up as knowledge. Combined with ideas, some of that is actually useful. Mix in experience, context, compassion, discipline, humor, tolerance, and humility, and perhaps knowledge becomes wisdom.
>
> (Turning Numbers into Knowledge, Jonathan G. Koomey, 2001, Analytics Press: Oakland, CA page 5, quoting Clifford Stoll.)

Today's companies have years' worth of customers' financial and other data saved in their various operational systems. A corporation's data resides in the different platforms and different operational units of the corporation. The question that begs an answer is if we have these mountains of data in our organizations, then why can't our executives use the data for strategic decision making? The problem we are facing is, how do we get from data to information? Consider, for example, this problem from the banking industry:

> Compare account activities from this year with account activities from the last 10 years.

In order to arrive at the answer, a DSS analyst must deal with lots of nonintegrated applications. For example, a bank may have different applications for loans and mortgages, separate saving and checking applications, and so on. Trying to get information from them on a daily basis is not practical, nor is it even possible. The major obstacle is that the historical data may be unavailable in those systems.

For example, deposits may have data for the last eighteen months, while the mortgage department has data for the last twenty-five years, but the loan department may have data for last two years, and so on. The systems in that environment are inadequate for supporting today's information needs. Two major obstacles are the lack of integration and the unavailability of historical data. That is why our corporations today are drowning in data and are facing an *information crisis*.

Business intelligence (BI) comes to rescue. Business intelligence, in essence, is the process of making better decisions, by turning data from the DWH into information, and then it into business knowledge. The knowledge obtained by means of it can pertain to customer needs, as well as insight into customer decision-making processes, industry trends, general economic trends and conditions, social trends, and other fields of interest.

Essentially, new data architecture had to be developed in order to support the new information requirements. The concept of *derived data* based on operational, *primitive data* is required in support of the DSS decision-making process.

The data needed for strategic decision making is the data from all those other systems, integrated and derived into an EDW, before it becomes suitable for any analysis. Therefore, EDW contains data suitable for strategic decision making for the business intelligence of an enterprise. This environment is completely separate from day-to-day operational environments. Only then can we think of applying BI, not as a product or an application, but as an architecture with a collection of integrated databases, decision-support applications and tools that provide the business community with easy access to business knowledge.

An important part of business intelligence is *analytics*. Analytics employs extensive use of data visualization, quantitative analysis, statistics, and explanatory and predictive modeling to support a fact-based, strategic decision-making process. Business intelligence includes all the processes involved in data access, reporting, and analytics.

Technically, analytics could be done using paper and pencils as it's the ideas, not the tools, that really count, but with today's state of IT technology and the amount of data, there is no need to do it that way. Instead, we should be looking for automated analytical tools. This book explores analytics using Excel, but there is a whole range of other statistical and predictive tools could be used for business intelligence. The important point that we are making here is that analytics is not about analytics information software, but the human interpretation of the data underneath it.

Some aspects of data visualization are illustrated in the data visualization section, as this is where the business intelligence industry will need to turn to succeed in the future. In today's business environment, in which our brains are bombarded with much competing information, decision makers don't have time searching for information; that is why a goal of the information visualization must be that the information is available at a glance.

Statistical regression not only produces predictions, but one that is accompanied by the confidence level of that prediction. Statistical data analysis provides the most powerful tools for understanding data, but the systems currently available for statistical analysis are based on an outdated computing model and have become much too complex to be used for this purpose. What we need is a simpler way of using these powerful analytical tools, which transparently presents data statistics visually in a much a simpler way. Statistical analysis uncovers hidden relationships among widely disparate kinds of information. The statistical analysis tools present the results of the analysis in a way that helps ensure that our intuitive visual understanding is commensurate with the mathematical statistics hidden under the surface. Thus, visual statistics ease and strengthen the way we understand data and they are used in information visualization of the abstract data underneath to amplify our cognition and create business insights.

Data Mining

The quiet statisticians have changed our world; not by discovering new facts or technical developments, but by changing the ways that we reason, experiment and form our opinions.

—Ian Hacking

There are many definitions for data mining (DM), but one from Gartner Group seems to be the most comprehensive:

> The process of discovering meaningful new correlations, patterns, and trends by sifting through large amounts of data stored in repositories and by using pattern recognition technologies as well as statistical and mathematical techniques.

DM is the science of extracting useful information from large data sets or databases. It is also known as knowledge discovery in databases (KDD). KDD is a multidisciplinary field covering information retrieval, machine learning pattern recognition, statistics, artificial intelligence expert systems, visualization, and databases. DWH is a useful source of data for exploration and data mining; as the data in the DWH is cleansed, integrated, and categorized historically, it becomes conducive for exploration and data mining.

Generally, data mining, sometimes called knowledge discovery, is the process of analyzing data from different perspectives, finding patterns, and summarizing them as useful business information—information that can be used to increase revenue, cuts costs, or both. Patterns, associations, or relationships among all this data can yield information. For example, analysis of retail point-of-sale transaction data can yield information on which products are selling and when. This information can be converted into *knowledge* about historical patterns and future trends. For example, a summary of information on retail supermarket sales can be analyzed in light of promotional efforts to predict the knowledge of consumer-buying behavior. This allows the producer or retailer to determine such things as which items are most susceptible to promotional efforts. Stores like Wal-Mart are constantly making an effort to have no excess inventory at all. What they have on the shelf is all they've got.

An important aspect of data mining is data visualization—the visual interpretation of complex relationships in multidimensional data where graphics are used to illustrate data relationships.

We could all testify to the growing gap between meanings of data at the time of data generation and our understanding of it at time of its use in the DWH project. As the volume of data increases inexorably, the proportion of people that understand it deceases alarmingly. DM is defined as process of discovering patterns in data. There is nothing new about this. People have always been seeking patterns—hunters look for patterns in animal behavior; farmers look at weather patterns; lovers seek patterns in their partners' responses, etc. In data mining, data is stored in computer memory and searches are automated. What is new is the ever growing number of opportunities for finding new patterns. With the swelling amount of data and with the abundance of computing power, data mining becomes our hope for elucidating hidden pattern in data. Intelligently analyzed data becomes strategic advantage which can lead to new insight bringing about competitive advantage.

The ultimate goal of data mining is prediction, and predictive data mining is the most common type of data mining and has the most direct application to business. Predictive analytics could be used to predict how customers would respond to a future promotion; in fraud detection, to identify which credit card number may be fraudulent in nature, etc. By learning from the abundant historical data, predictive analytics provides the business value beyond standard business reports and sales forecasts; it provides actionable predictions for each customer.

The idea of learning form data has been around for a long time. But despite these nice promises the goal of DM, as most of the gold rush of the past, has been to *mind the miners*. The largest profit has been in selling tools to the miners. DM, too, has been used as a means to sell computer hardware and software. A quote of Chuck Dickens, former director of computing at SLAC:

> Every time computing power increases by a factor of ten, we should totally rethink how and what we compute.

Since all currently DM tools have been invented, computing power and size of data has increased by a several orders of magnitude. So we should soon look forward to much brighter future for the new DM methodology[3].

---

[3]   Jerome H. Friedman: *Data Mining and statistics: What is the Connection?*

Quality Assurance Story

It is not enough that the top management commit itself for life to quality and productivity. They must know what it is that they are committed to—that is, what they must do. Those obligations cannot be delegated. Support is not enough; action is required.

—Dr. W. Edwards Deming

One spring day, after the winter winds had died down, after long cloudy night skies finally gave way to the stars and the moon, the most extraordinary thing occurred.

The king woke up from the in his dream in which he heard the voice telling him the most amazing story about this great land that had elephants. Next morning, he summoned his three wise advisors. He told them a story that he had heard in his dreams, about a great land that had elephants. In spite of the fact that he had not seen an elephant and did not know what kind of animal an elephant is, he was sure that this was just what he needed to save his country from its poverty.

The king's advisors had not seen an elephant before either, but they were sure they could get him some. Since they did not know what an elephant is, they travelled to this distant land to see it. They arrived in the great land in the middle of the night. They were tired from their long journey, but they insisted on seeing an elephant as soon as possible, so that they could go home and give their king one or even more of these elephants.

They were taken to the Elephant House in the middle of the dark, moonless night. The first advisor entered the dark room and began to feel around for an elephant. He touched the elephant's leg. He immediately declared it's a tree! The second advisor touched its tail and declared it's a snake! The third advisor bumped into the elephant's side, and it felt exactly like a wall. He said to himself, it must be a wall!

The three advisors went home and declared themselves to be the *Elephant Masters*. Soon, they started roaming the land, giving expensive advice on how a person could find an elephant on their own.

The king, disappointed that he did not get his elephant, on which he had placed great hopes for his land, ordered the three advisors to be locked in the *Elephant Institute* every day from dawn till dusk, until they gave him and the people of his land an elephant, or until the age of sixty-five.

Quality assurance is not a thing that one can somehow get. Neither is it a destination that one arrives at. Quality assurance is the commitment to making a long journey! It is the road less traveled; the road that twists and turns, and no two directions are ever the same. There is no absolute definition of quality assurance, as the processes can always be improved and taken to the next level.

To pursue quality assurance is to reject the complacency of thinking that we have finally *gotten it perfect* because there is no final end; it's a moving target—a journey—as our joy is found not in finishing an activity, but in doing it.

Ithaca is the island in Greece that Odysseus had so much trouble returning to, in Homer's *Odyssey*. Greek poet Constantine Kavafis wrote a poem called Ithaca in 1911 explaining it. Here is a part of it:

> Keep Ithaca always in your mind.
> Arriving there is what you're destined for.
> But don't hurry the journey at all.
> Better if it lasts for years,
> so you're old by the time you reach the island,
> wealthy with all you've gained on the way,
> not expecting Ithaca to make you rich.
>
> Ithaca gave you the marvelous journey.
> Without her you wouldn't have set out.
> She has nothing left to give you now.
> And if you find her poor, Ithaca won't have fooled you.
> Wise as you will have become, so full of experience,
> you'll have understood by then what these Ithacas mean . . .

Defining quality is hard because everybody seems to have their own preconceived idea of what quality means. But for the most part, they disagree amongst each other. It is like in the above parable, frequently told in management, about three wise men that came across an elephant and could not agree on what an elephant is. Quality, too, may mean different things to different people.

Consequently, when two people are discussing quality, they often are talking about different things. A *customer-oriented user* may be looking for ease of use or quick transaction time. A more technical *knowledge user* may be looking for a rich set of functions. A project manager may be looking for all of the above and for delivering it on time and within the budget.

The quality of DWH application is an elusive aspect of it, not because it is hard to achieve (once we agree what it is), but because it is difficult to describe.

In the context of this book, we do not see quality according to the classic definition by Gause and Weinberg: A problem can be defined as the difference between things as perceived and things desired.

Instead, we propose the notion that quality is not an attribute or a feature of a product, but rather a *relationship between that product and a stakeholder*. More specifically, the relationship between the software quality and the organization that produces the products is explored. This is a multifaceted relationship, dependent on many factors—business strategy, business culture, available talents, processes that produce the products, etc. We are purposely using a slightly more inclusive view of quality, as quality, especially in DWH application, is more elusive and highly unpredictable.

The conceptual framework for software quality was built on the pioneering work of Dr. W. E. Deming. It was his leadership in statistical quality control that propelled Japanese industries (automobile, shipbuilding, electronics, and others) to world prominence back in the 1950s and 1960s. More recently, the performance of the Japanese software industry is showing that these concepts of statistical process control are as applicable to software development as they are to producing consumer goods like cameras or television sets. The basic principle of statistical control is measurement. It is said that *what gets measured gets improved*, or as Lord Kelvin said one century ago:

> When you can measure what you are speaking about, and express it in numbers, you know something about it; but when you cannot measure it, when you cannot express it in numbers, your knowledge is of a meagre and unsatisfactory kind; it may be the beginning of knowledge, but you have scarcely in your thoughts advanced in the stage of science.

It is the function of a QA Team to plan a systematic pattern of actions that will provide the confidence that all the stakeholders' expectations are met and at the same time, ensuring that DWH application project standards and procedure are maintained throughout all the phases of the project and ensuring the objective of QA in preventing defects from being built into products!

Consequently, quality is built into the products at all stages of the SDLC, but when it comes to testing, it is too late! Nevertheless, the goal of QA is not QA activities and processes, but the awareness of everybody who works around these QA processes that the goal of QA is to deliver a tool to business users, which will enable them to better steer the ship through the stormy sea in front of us.

From the quality assurance point of view, it's not how well one can measure the quality of the product, even though this is an important activity, but this will not achieve the desired product quality. Measuring product quality is just a feedback—a learning opportunity—on how effective is the software quality program that produced the product!

Quality is affected by many factors; however, only those that are actually producing the product build the quality into it! QA actions and methodology must be focused on integrating the efforts of the many and the few, bringing them together to achieve quality.

The software quality program and overall approach in achieving the desired quality are laid in and are an integral part of the Test Strategy document. The activities necessary to achieve the quality objectives are described in it:

- Establish clear, objective, and verifiable requirements for the product quality.
- Implement and enforce methodologies, processes, and procedures to support the quality objective.
- Implement validation and verification methodology, processes and procedures to evaluate the quality objective throughout all the phases of the SDLC.

This book concentrates on differences in QA processes between the DWH application and classic software application projects. Software quality methodology, such as the CMMI (capability maturity model integrated) is still required but not sufficient for DWH application projects. The fundamental notion in CMMI is the management premise that

> The quality of the system or product is highly influenced by the quality of processes used to develop and maintain it.

CMMI provides a comprehensive description in different stages of maturity and the conditions that determine where one is and where one can hope to be in order to grow, thus turning the corner from chaotic software development to a controlled management process.

While it may seem trivial to define the current state of software development processes in an organization, it is not. One way to assess the software capability of an organization is to watch what it does in a crisis. That is when the good practices are most critical and when guidance is required.

If the processes in use by the software development project are not defined and organized well, the quality of the software product is neither predictable nor repeatable. The dependence of software quality and processes that are producing the software products is the basic assumption of the Software Engineering Institute (SEI) in their CMMI model for software development. Five levels of maturity are described by the CMMI:

Level	Characteristics	Key Problem Area
Initial	Processes are ad hoc and chaotic	Project management Project Planning Configuration management Software quality assurance
Managed—Repeatable	Process is managed in accordance with agreed metrics	Training Technical practices: - reviews, testing, etc. Process practices: - standards, process groups, etc.
Defined	Processes are well characterized and understood, and are described in standards, procedures, tools, and methods	Process measurement Process analyses Quantitative quality plans
Quantitatively managed	Quantitative objectives for quality and process performance are established and used as criteria in managing processes	Problem causal analyses Problem prevention
Optimizing	Process management includes deliberate process optimization/ improvement	Process automation

Low maturity software organizations spend most of their time and energy fighting fires in trying to contain a high volume of changes or recovering from late-discovered defects found in production. The same defect is rediscovered many times, and many solutions are reinvented as a management process is nonexistent. Any organization—where the heroic efforts of a lone cowboy, rather than an organized QA process of defect prevention are more popular with the management—is also risky place to work. When a crisis hits, their solution is the technical wizard; unfortunately, there is often no way to recover. As Dr. F.P. Brook says, "There is no silver bullet."[4]

---

[4]   Dr. F.P. Brooks, *"The Mythical Man-Month: Essays on Software Engineering"*—
       The classic book on the human elements of software engineering

The solution is systematic project management, where work is estimated, planned, and managed, and where quality assurance is an independent team, charged with the responsibility of assuring that all the processes are properly performed.

For an excellent discussion of the CMMI process, we refer interested readers to the works [Ref. 41] of Watts S. Humphrey, also known as the *father of software quality*, a fellow and a research scientist at the software process program of the SEI at Carnegie Mellon University in Pittsburgh. He wrote the first version of the SEI's CMMI for software in 1987.

Some of Brooks' insights and generalizations are:

The Mythical Man-Month: Assigning more programmers to a project running behind schedule may make it even more late.

The Second-System Effect: The second system an engineer designs is the most bloated system she will EVER design.

Conceptual Integrity: To retain conceptual integrity and thereby user-friendliness, a system must have a single architect (or a small system architecture team), completely separate from the implementation team.

The Manual: The chief architect should produce detailed written specifications for the system in the form of a manual, which leaves no ambiguities about any part of the system and completely specifies the external specifications of the system i.e. what the user sees.

Pilot Plant: When designing a new kind of system, a team should factor in the fact that they will have to throw away the first system that is built since this first system will teach them how to build the system. The system will then be completely redesigned using the newly acquired insights during building of the first system. This second system will be smarter and should be the one delivered to the customer.

Formal Documents: Every project manager must create a roadmap in the form of formal documents which specifies milestones precisely and things like who is going to do what and when and at what cost.

Communication: In order to avoid disaster, all the teams working on a project, such as the architecture and implementation teams, should stay in contact with each other in as many ways as possible and not guess or assume anything about the other. Ask whenever there's a doubt. NEVER assume anything.

Code Freeze and System Versioning: No customer ever fully knows what she wants from the system she wants you to build. As the system begins to come to life, and the customer interacts with it, she understands more and more what she really wants from the system and consequently asks for changes. These changes should of course be accommodated but only up to a certain date, after which the code is frozen. All requests for more changes will have to wait until the NEXT version of the system. If you keep making changes to the system endlessly, it may NEVER get finished.

Specialized Tools: Every team should have a designated tool maker who makes tools for the entire team, instead of all individuals developing and using their private tools that no one else understands.

Humphrey also developed the personal software process (PSP) and the team software process (TSP). His work was based on Dr. W. Edward Deming's principles of statistical process quality control, summarized in Deming's fourteen management principles [Ref. 42] and Humphrey's strong practical background at IBM, where he worked from 1959 to 1986 as a director of programming quality and processes, including on IBM's most successful OS/360 project. We are associated with neither SEI nor Mr. Humphrey, but the above credentials are strong reason for recommending his work.

A fundamental problem, however, with the CMMI or any other QA software engineering methodology in developing DWH applications, or the most important part of it—mainly business intelligence—is that BI is more in the realm of art than it is in engineering. The more your try to control it, the more it disappears!

Before continuing with the descriptions of various QA processes, it is important to stress that everyone on the project communicate using the same terminology; otherwise, a project may end up like the ancient Tower of Babylon:

> Up until that point in the Bible, the whole world had one language—one common speech for all people. However, the people of the earth became skilled in construction and decided to build a tower that would reach to heaven. God came to see the tower they were building. He perceived their intentions, and in His infinite wisdom, He knew this "stairway to heaven" would only lead the people away from God. He noted the powerful force within their unity of purpose. As a result, God confused their language, causing them to speak many different languages so they would not understand each other. By doing this, God thwarted their plans. That is why the Tower of Babylon collapsed.

The story is inspired by Bible: Genesis 11:1 and 11:7.

Fig. 5.1 Tower of Babylon

Software Testing vs. Software QA

Let's start by clarifying the differences and distinctions between software testing and software QA. The two terms have often mistakenly been used synonymously, but these are two distinct processes, which, however, are complementary and, if properly employed, synergetic. They have different goals and require a different skill set. QA is a management/leadership activity responsible for setting up and enforcing processes and methodologies in order to prevent defects. Defect prevention is a powerful technique for improving software process. The general idea is to track every defect and periodically perform causal analyses. Depending on the project and the organization, there are number of standards (CMMI, ISO, IEEE, etc.) that can be deployed in the software development process of DWH application.

Testing, on the other hand, is a technical activity responsible for finding defects—all of them and sooner rather than later. If the QA describes *what needs to be done*, testing describes *how it is going to be done*. Testing is the subject of another chapter in this book.

Software failure is the inadequacy of the system to execute a required function within specified time. It is manifested by incorrect output, abnormal program termination, or system crash. Software fault is caused by incorrect or missing code.

A bug is a software fault too. It was first used to describe a problem when the moth got wedged in the computer relay causing the program to halt. A bug is corrupting euphemism. A bug should not be called *a bug*, as it metaphorically blames the cause of the fault to some evil forces beyond the programmer control. It is much more honest to call it an error or a defect, as it is created by the programmer's action. This slightly changes the perspective on the program that does not work and places the responsibility where it belongs.

Validation versus Verification

Other terms that are also often mixed up are validation versus verification (V and V). Definitions as used in this book are presented here.

The purpose of validation and verification is to establish with confidence that the data warehouse is fit for its intended use. However, this does not guarantee that the warehouse application is completely free of defects.

The validation and verification testing processes must be integral part of every phase in the entire software development life cycle (SDLC). They must be applied at each stage in the SDLC, with principal objectives including:

- The continuous assessment of whether or not the DWH application is useful and useable in an operational situation by business users.
- The prevention and discovery of defects in the DWH application.

Validation

First, we do the validation to answer the question, are we building the right product? In other words, does the software do what the user actually requires?

Business users of the data warehouse application must be provided with all the objects, such as use cases, data flow diagrams, VISIO diagrams, and mock-up screens and reports based on near-production data, etc. This enables them to fully understand the solution and ultimately validate that the EDW under development complies with their requirements, performs functions for which it is intended, and meets the organization's goals and user needs. A typical reaction when the user sees the application in production for the first time is:

Yes, this is nice, but now that I see how it works, wouldn't it be better if . . .

Avoiding the *yes, but syndrome* is the reason for having the user involved in the validation process. The root of the *yes, but syndrome* is the inherent intangibility of the intellectual process. The *yes, but syndrome* can be avoided by providing users with all the intermediate objects and keeping them involved in the validation and decision-making process. It is important to keep in mind that the user being *busy* is not an acceptable excuse, since it's always cheaper to build the application right the first time, than to do it all over again.

Verification

Only when we know what we are going to build do we proceed with the verification. Are we building the product right?—this is a question that we must answer during the verification process. Or translated into software parlance, does the software conform to its validated specification?

Verification is performed during development on all key deliverables, like walkthroughs, reviews, inspections, and testing. The goal of verification is to demonstrate the consistency, completeness, and correctness of the DWH application at each stage and between each stage of the software development life cycle.

Verification ensures the adherence to validated user requirements, the goal which is determining if the system is consistent with validated requirements and if it adheres to standards and performs the selected functions correctly.

Both effective and efficient software processes are critical to successful EDW implementation, but their merit can only be determined in the context of the business needs of the particular organization.

Classical software development life cycle (SDLC) revolves around the *Waterfall* or *V* model.

V and V are depicted and marked as appropriate to that model in the picture below.

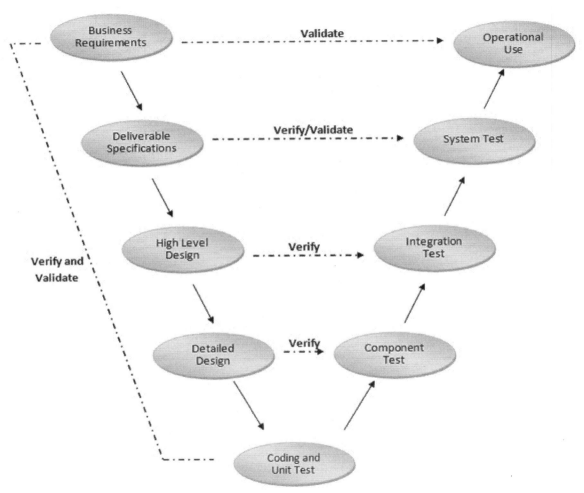

Fig. 5.2 Waterfall SDLC

Testing at Every Stage of the DWH Development Cycle

Everything should be as simple as possible, but not simpler.

—Albert Einstein

Complexity of data is being recognized more and more as the key characteristic of the world we live in. The size of data compounds the problem of complexity by masking the hidden patterns upon which the complexity is built. Sometimes, data complexity is built upon the simple pattern of data redundancy. Seldom are we interested in uncovering all the patterns and relationships of data, but we are mostly interested in the few that are required to answer questions posed by businesses.

The central task of a BI data analyst is to show that complexity is but a mask for simplicity by finding hidden patterns in mountains of data. For when the wonder of complexity is explained, using aesthetics of art and mathematics by discovering concealed patterns, a new wonder arises of how complexity was woven out of simplicity. It creates the opportunity for data analysts to first find and analyze these patterns and then use them, by virtue of syntheses, in crafting a convincing story to form the new complexity. These uncovered patterns and relationships of data are presented to users as cognitive objects (graphs and reports) and can be thought of as an *interface* between the user's mind and the external data. This has never been done before in the short history of software development.

The development of DWH application evolves at the junction of two worlds—the IT world and the business world. DWH application is not a project, but rather a program that goes on with no endpoint or finished deliverable. Yes, there are concrete business requirements, but to plan it as typical software deliverable that is *finished* at any point in time should not be an expectation. DWH application is a project that should live as long as the business needs information. It does not follow any traditional software life cycle of birth, life, and death.

The DWH application continues to evolve over time, unlike a software development project that is completed when the project is deployed. Software application testing methods focus mostly on testing transaction-oriented systems. These involve code testing, while DWH application testing is different in many respects. Yes, classic software testing is an important part of DWH application testing, but it is not the only part.

The DWH application development process is highly iterative and without a finite end, therefore requiring continuous testing. The scope of testing goes beyond just software testing. It starts with testing the IT alignment with business goals and ends by testing delivered business value.

Business requirements for the DWH application projects are usually fuzzy, indefinite, and incomplete, less stable and more prone to change. The potential for additional requirements that may affect the design in the early development cycle is very high. This is because when the user is presented with the solution, he will recognize the potential of the technology. That is why it is very important to have the business user involved in the project from the beginning.

While software application testing has the goal of ensuring the functional correctness of the code, DWH application testing focuses on the business value of the information derived by the delivered data. Information quality goes beyond data quality. Special consideration should be given to data quality, and perhaps the most important aspect of data quality—data appropriateness.

We often measure and collect information about those things that are easy to count. This may give us a false sense of security, but by ignoring those intangible (hard to measure) factors, we may be missing the forest for the tree.

For instance, it is much easier to measure the quality of a product than its impact on the quality of life. We measure what we can and ignore the res. By ignoring what is left, we may be ignoring a critical part of a typical company that is based on intangible assets.

Consequently, what gets measured gets improved, because we value what we can measure. We may be doing a superb job on something of little consequence and leaving out the most critical part. This is why we say that before we start being efficient, we first must make sure we are effective.

The foundation of software application testing is found in business requirements and system designs. The requirements for building and testing DWH applications are much more elusive. The objective of software application testing is to verify the correctness of the program code and that it does not fail (and if it does fail, it fails gracefully). This must be done before the application is deployed in production. DWH application testing is directed at data and information testing and it goes on (should go on) long after the DWH is deployed.

Implementing a DWH application initiative is not a project, but a long-term commitment to implementing continuously improving business intelligence practices. Consequently, a DWH application development process is highly iterative and without a finite end, therefore requiring continuous testing.

Data warehousing applications use a distinctly different SDLC model, which illustrated in the diagram below. DWH application development is a new paradigm, and it requires a new SDLC. We are proposing the model that we have been using successfully—Perpetual (never ending) Iterative SDLC. The name is fair to our businesses; they should also be made aware that these DWH projects never end. At the heart of this process is validation, as can be seen in the picture below.

Validation at every step of the cycle.

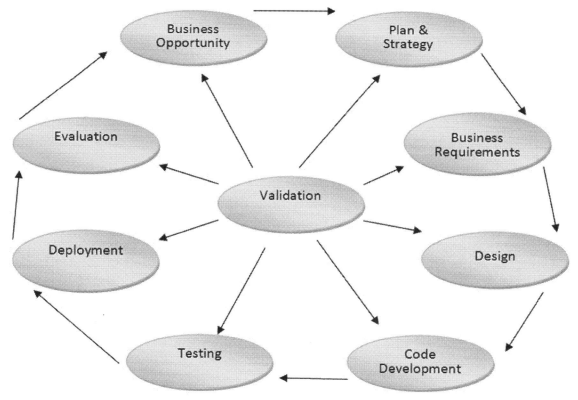

Fig. 6.3 DWH SDLC Model

The differences discussed above dictate changes in the:

1. Structure of the QA and development teams,
2. Software development life cycle, and
3. Nature of testing.

How many times have we heard business people telling us

> I don't need all those reports—just give me the numbers that I need to find out what's going on.

It is the IT people that determine what those numbers that the business user needs are. Even though these are business professionals who have gone to different universities than the IT professionals don't know anything about their day-to-day work, they still want to provide them with reports that they need in order to do their work.

Programming is intellectually demanding and all-absorbing work that dominates over all other considerations, including alien thought processes concerning users. Programs want to do a good job, as everybody else does. So what do programmers do without full requirements? If, for example, programmers are asked to create an accounting system report, they go and read some accounting books and magazines, find some fancy graphs and reports, and then improve on them—*improve* on something they do not know anything about (!) by adding bells and whistles to dazzle (intimidate) the business users. Correct reports have been delivered, except that no one knows if they are usable, and the project is completed and closed. But without knowing if the project has delivered the ultimate purpose of business intelligence, the conclusion that cognitive objects that stimulate cognitive thinking cannot be made.

The quote below is very appropriate for the state of DWH application projects today.

> High-tech companies—in an effort to improve their products—are merely adding complicating and unwanted features to them. Because the broken process cannot solve the problem of bad products, but can only add new functions, that is what vendors do . . . The high-tech industry has inadvertently put programmers and engineers in charge, so their hard-to-use engineering culture dominates . . . Programmers aren't evil. They work hard to make their software easy to use. Unfortunately, their frame of reference is themselves, so they only make it easy to use for other software engineers, not for normal human beings . . . While we let our products frustrate, cost, confuse, irritate, and kill us, we are not taking advantage of the real promise of software-based products: to be the most human and powerful and pleasurable creations ever

imagined . . . All it requires is the judicious partnering of interaction design with programming.

(*The Inmates Are Running the Asylum*, Alan Cooper, 1999, SAMS Publishing: Indianapolis, Indiana, page 8, 15, and 17)

If you ask, as we have, any QA professional who has delivered DWH application to business:

How do you know the reports you have delivered are useful? You know the reports are correct, but are they useful to your users?

Typical responses would be, "I know because our users are not reporting any defects!" In other words, reports are correct, but probably useless. This is not the exception; it's the norm in the BI industry.

The traditional software vendors that are providing correct and effective data warehouse solutions are not able to move into the realm of visual analytics, which is where the information can be analyzed, explored, and used to predict the future. Success in the BI industry has come from a few nontraditional vendors like Tableau, TIBCO Spotfire (spin-offs of universities and research), and companies like SAS with a long history of work in statistics.

Is it any wonder that the *yes, but syndrome* keeps rearing its ugly head? That is why some authors are suggesting that the first delivery of BI reports to business users is a throwaway and a good place to start collecting real business requirements. We disagree, as the next iteration will suffer from the same disease.

Much more than that has to change, as this is not where the problem is, because the subsequent iteration will be a throwaway as well. We know how to collect, clean, transform, and store data, but we don't know how to deliver information that can be analyzed and used to predict the future. This is where our QA methodology differs from all others.

Quality Assurance Strategy for DWH Applications

If you can't describe what you are doing as a process, you don't know what you're doing.

—Dr. W. Edwards Deming

We begin with the end goal in mind, but the goal without a plan to achieve it is just a dream. That is why we create a plan which prescribes *what needs to be done* in order to achieve the end goal of successfully deploying the DWH application. The plan is typically laid out in a document called the test strategy or master test plan. The *strategy* is the term we prefer to use in this book, as the term metaphorically relates to the word *war*, which conceptually relates to the brain, the importance of this document.

Testing strategy for DWH application, first of all, must be aligned to business strategy. The very first priority of the strategy is mapping the top business questions into a problem definition and to achievable testing strategy objectives. But before starting to resolve the problem, the problem has to be defined first. More specifically, there must be processes to ascertain what the key business question is. Only then we can start mapping the business problem to data visualization solutions.

Identifying the business problem is a team effort, between business users (domain expert), data analyst, and business analyst. QA must be privy to this effort in order to be able to map the successful testing strategy. As the requirements for the success of the DWH application are vague and abstract, and very difficult to visualize even by business user, some kind of prototype must be employed for continuous verification of the proof-of-concept. Success criteria must include terms such as return on investment (ROI), profit, and proposed matrix to track those.

Key performance indicators (KPIs) should be employed and agreed upon by all the stakeholders. Most of the time, KPIs are confused with performance target values. Use of KPIs in business intelligence is to assess the present state of the business and to prescribe a course of action. The KPIs are dependent on the nature of the organization and the organization's goals and mission. KPI are tactical tools used for *mid-course evaluation and correction*. They help an organization to measure progress toward their business goals, especially toward difficult-to-quantify knowledge-based goals. KPIs can be expressed in many ways such as cost, quality, time, units of service, etc. These can be expressed in time, volume, money. KPIs can be measured in absolute numbers, ratios, or process improvement units.

Strategy defines not only all the processes, but which processes, must be performed at which point of the project SDLC, in which environment and by whom, in order for the DWH application to deliver expected business value. The necessary ingredient of the testing strategy is to define what success means. The success of the project is measured by its success criteria (how do we know we have succeeded?). The definition for success of the organization must be directly related to success of the testing strategy.

Elements of the strategy document are lists all the resources and tools—type of tests, design and testing documentation reviews, project assumptions related to QA, risk identification and risk mitigation, timelines and processes—required to achieve the end goal of delivering quality product, on schedule and within budget, and to the expectation of the business user.

Testing is the most critical element of the strategy. There are several types of test, but from our experience, four have been the most valuable in DWH application testing—unit tests, functional tests, integration tests, and regression tests. These are explained further in the later part of this chapter. Automated testing, a special category of regression test, is described and demonstrated in the last chapter of the book.

Risk and risk mitigation are also part of the strategy document. At the early stage of a project, all the assumptions must be validated. For an example, one of the project assumptions may be that the company has all the data it needs to populate DWH. If, however, the firm wants to build the data warehouse, but it is still using double-entry accounting system, invented five hundred years ago, it is hard to believe that this company may have data it needs for the DWH application.

Double-entry accounting system was invented by Fra Lucia Pacioli five hundred years ago, in Venice Italy and documented in his book *Everything about Arithmetic, Geometry and Proportions*. Double-entry accounting, based on tangible assets, was good then, but it is not sufficient anymore today, where the market value of many modern companies (i.e., Google, Microsoft, etc.) is mostly based on intangible assets. It is unlikely that the company using double-entry accounting system would have data required for BI.

The greatest risk in developing DWH project is delivering DWH application that the business users cannot use. This book presents the approach to guard against that risk.

The same QA methodology that is applied to any software systems can also be applied to DWH applications quality assurance. The QA testing cycles are the same; we still must execute all the regular testing, including, unit, integration, systems, performance (stress) and user acceptance testing (UAT). However this is where the similarities between software application and DWH application testing strategy end.

Validate, Validate, Validate!

As already mentioned, we have to know what it is we are building, before we start building it. Early and continuous involvement in testing provides the opportunity for early error detection and prevents migration of errors from requirement and specification to design, and thence from design to being built into code. The earlier in SDLC the errors are uncovered the easier and less costly they are to fix.

Ours is the era of predominantly visual perception. Messages are received through the eyes first. The eyesight needs the shortest possible time, in comparison with other channels, to make impression on the brain. Vital, quantitative business information is communicated to the brain in form of tables and graphs. G graphs are being used today, in every project and on every possible occasion, to convey messages and stores effectively; sometimes, these are simple line or pie charts to present complicated correlation within the data.

One of the functions of QA is to verify the usability of the delivered products. User requirements in DWH applications are more elusive than ever before. But QA is also responsible for validating these requirements and convey them unambiguously to the development teams.

The QA methodology, we are about to lay down before you, is our original approach, and it has been proven in our practice with many of our clients. We know that as we are often able to see that a client has produced tangible success as a result of our services.

We have plenty of empirical evidence of the beneficial outcomes of this approach, but we are not allowed to share them publicly. In any case, we hope to be able to convince our readers, by the end of this chapter, that this QA methodology will shorten your development life cycle and diminish the possibility of errors. At least, it is quicker and less expensive to do it right the first time, instead of repeating the entire effort all over again. We are convinced that this approach will become ubiquitous in future. Two processes are absolutely essential and at the heart of this approach:

- Cognitive Data Visualization, and
- Prototyping.

Just one more thing [as Steve Jobs used to say] Have we said that this methodology will not work unless the business assumes the responsibility of assuring the usability of the delivered DWH application? As the domain expert knowledge is within business community, the responsibility of assuring the usability of the delivered BI application must clearly be placed on the business user's shoulders.

Data Visualization

The greatest value of a picture is when it forces us to notice what we never expected to see.

—John W. Tukey, *Exploratory Data Analysis*

Data Visualization (DataViz) is the graphical presentation of multidimensional data with the goal of carrying out analyses and exploration of data to identify patterns, associations, trends, and so on. A graphical presentation of quantitative information always involves some aspect of relationships among the values. Good data visualization helps users explore and understand the patterns and trends in data and communicate that understanding to business users to help them make robust decisions by providing an effective representation of the underlying data being presented.

For now, DataViz practice is more in the realm of art than it is engineering, or even a scientific discipline. Most people who are responsible for analyzing data have never been trained to do this. Knowing how to use Excel or any other software that can be used to analyze data is not the same as knowing how to make sense of data.

DataViz practice needs to define a body of knowledge and a set of rules and practices before questions, such as the ones below, can be answered with certainty:

- How to select the best graph to visually display a data set underneath.
- How to interpret a scatter plot or box-and-whiskers graph and when to use them.
- How to create the charts and graphs not there to impress the users but to answer their questions.
- How to prioritize information with plots of trend patterns.

The basic difficulty that DataViz practitioners are facing is that sufficient theories do not exist to describe and predict important phenomena of information visualization. Most fundamental questions are largely unanswered. Why does a certain visualization technique work better than another? How can we make a visualization tool better? How do viewers gain insights? The existing theoretical work is far from sufficient to help in answering these questions. The DataViz discipline is still in its inception, and the biggest barrier to forming reasonable theories is a lack of empirical evidence.

Tables are used to present detailed (or summarized) levels of a data set, but just by looking at the data in the table, it's impossible to ascertain any trends or relationships. It is the graphs that enable visual communication of information about variables in the data sets, such as trends, ranges, relationships, frequency distribution, comparisons, etc.

The goal of a good graph is to display and communicate patterns and relationships of the underlying data with simple clarity. Additionally, looking at multiple graphs simultaneously (case for dashboard) can offer a new insight about data sets.

A DataViz practice was invented by Edward Tufte [Ref. 6], and he still rains in it. Stephen Few [Ref. 7] followed the work of the great maser. Stephen Few [Ref. 8] is seen as an illuminator of Tufte's ideas as they pertain to DataViz applied to business. Edward Tufte and Stephen Few are often cited together.

The purpose of analytical displays of evidence is to assist in thinking. Consequently, when constructing displays of evidence, the question is, *what are the thinking tasks that these displays are supposed to serve?*

The essential claim of Tufte's book, *Beautiful Evidences*, is that effective analytic design entails turning thinking principal into seeing principals. So if the thinking task is understanding causality, the task calls for design principle, *show causality*. If a thinking task is to answer a question and compare it with alternatives, the design principle is to *show comparisons*.

In the excellent book *The Visual Display of Quantitative Information*, Edward Tufte describes the value that good data graphics must have[5]. Good graphics must "above all else, show the data" and should

- help the audience think about the important message(s) from the data, rather than about methodology (graphic design, the technology of graphic production, etc.), or something else;
- avoid distorting what the data have to say;
- present many numbers in a small space—but also emphasize the important numbers
- Make large data sets coherent, and encourage the audience to compare different pieces of data; and
- reveal the data at several levels of detail, from a broad overview to the fine structure.

---

[5]   Tufte (2001): *"The Visual Display of Quantitative Information"*

Even before Tufte and Few, authors like William S. Cleveland [Ref. 9] had explored DataViz with paper and pencil. Messages arriving to our brains are predominantly (70 percent) delivered though our eyes. One of the leading experts in visual communication, Dr. Colin Ware, explains:

> Why should we be interested in visualization? Because the human visual system is a pattern seeker of enormous power and subtlety. The eye and the visual cortex of the brain form a massively parallel processor that provides the highest-bandwidth channel into human cognitive centers. At higher levels of processing, perception and cognition are closely interrelated, which is the reason why the words "understanding" and "seeing" are synonymous. However, the visual system has its own rules. We can easily see patterns presented in certain ways, but if they are presented in other ways, they become invisible . . . The more general point is that when data is presented in certain ways, the patterns can be readily perceived. If we can understand how perception works, our knowledge can be translated into rules for displaying information. Following perception based rules, we can present our data in such a way that the important and informative patterns stand out. If we disobey the rules, our data will be incomprehensible or misleading. (Ware, Colin, *Information Visualization*, Second Edition, Morgan Kaufmann Publishers, 2004)

At this point in time, there are no well-rounded data visualization theories. There are some general principles, and from there, a set of best practices can be derived. In the next section some basic data visualization principles using Excel charting techniques are demonstrated. The Excel charts can be effectively used to create powerful data visualization objects. But we emphasize that to becoming a master of the trade, it's not the tool, but it's the knowledge how to get the most out of the data, deeper understanding of what charts should be used and when it is required, and how to analyze and visually communicate quantitative information.

"Less Is More"

Simplicity is the ultimate sophistication.

—Leonardo da Vinci

This is the basic principle in DataViz, and it can be traced back to Edward Tufte.

In his book, *The Visual Display of Quantitative Information* [Ref. 6], Tufte refers to these concepts as data-ink ratio, data-density, or chartjunk.

"Data graphics should draw the viewer's attention to the scene and substance of the data," says Tufte.

"The purpose of visualization is insight, not [pretty] pictures," says Shneiderman.

Tufte brilliantly transposes this concept from Ludwig Mies van der Rohe's minimalism to the field of data visualization. The minimalistic approach is applicable to any field of human endeavor and, most certainly, to human communication.

We are quoting a brilliant illumination of Tufte's data-ink ratio, a basic principle of data visualization by Stephen Few. In his book, *Show Me the Numbers: Designing Tables and Graphs to Enlighten*, he writes,

> The process of reducing the non-data ink involves two steps:
>
> 1. Subtract unnecessary non-data ink
> 2. De-emphasize and regularize the remaining non-data ink
>
> Subtract unnecessary non-data ink
>
> The process of subtracting unnecessary non-data ink involves asking the following question about each visual component: "Would the data suffer any loss of meaning impact if this were eliminated?" If the

answer is no, then get rid of it. Resist the temptation to keep things just because they're cute or because you worked so hard to create them. If they don't support the message, they don't serve so hard to create them. If they don't support the message, they don't serve the purpose of communication. As the author Antoine de Saint-Exupery suggests: "In anything at all, perfection is finally attained not when there is no longer anything to add, but when there is no longer anything to take away." (*Show Me the Numbers: Designing Tables and Graphs to Enlighten*, page 118 under heading: "Reduce the Non-Data Ink")

The notion that *nothing* is an important *something* represents Tufte's concept of reducing data-ink ratio. Increasing the amount of blank space is compensated by enhancing the importance of what is left. Indeed, when there is less, we appreciate more of what is left. In other words, the graphic should focus on the message(s) for the audience, and all visual clutter should be kept to a minimum.

What's Wrong with Pie Charts?

Pie charts are perhaps the most ubiquitous chart type; they can be found in newspapers, business reports and many other places. In spite of their widespread use, many data visualization writers like Edward Tufte, Stephen Few and Naomi Robins do not recommend the use of pie charts.

In *The Visual Display of Quantitative Information* [Ref. 6], Tufte writes, "The only worse design than a pie chart is several of them."

In his highly regarded book, *Show me the Numbers*, Stephen Few says: "I don't use pie charts, and I strongly recommend that you abandon them as well."

Instead, Dot Plots or simple Bar charts are recommended as an excellent alternative to pie charts, because they show data positions along a common scale, rather than relying on pie chart angles.

Here is the proof, by exhibiting an example, for the above statements:

Description	Surface [km²]	Area surface area in percent [%]
Saltwater	352,103,700	69.03
Freshwater	9,028,300	1.77
Farming land	44,682,307	8.76

Mountains	29,788,205	5.84
Snow-covered land	29,788,205	5.84
Dry land	29,788,205 km^2	5.84
Land without topsoil	14,894,102	2.92

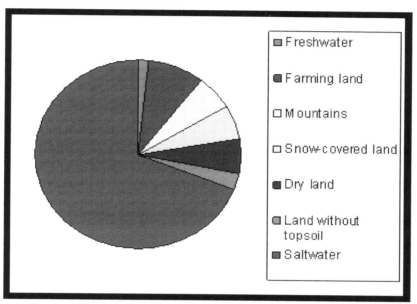

Fig 6.4 Pie Chart

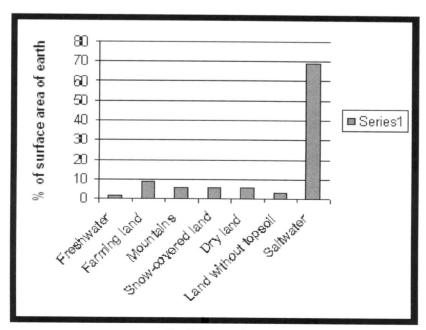

Fig 6.5 Bar chart

The simple rule for use of the pie charts is if it has more than three slices, do not use pie charts. Unless it's one of these:

OR →

Fig 6.6 Fig 6.7

It is not with lightheartedness that we are displaying the above symbols of German *Gestalt* philosophy, but to present the functional beauty of the graph principle, another important attribute of data visualization.

The German word for design is *gestaltung*; this term originated in the famous Bauhaus school founded in 1919. The basic principle of *gestalt* is to aspire to find the conceptual fit that resonates with the mind. The goal of *gestalt* practitioners has been the relentless pursuit to find the most appropriate *gestalt* that benefits a need.

If the objects we are creating are going to inspire conceptual thinking and business insight (i.e., Business Intelligence), they must have aesthetic appeal. They must be in harmony with our user's minds.

Representing numbers

One of the oldest methods of representing numerical quantity is by tally marks. When counting a number of items, we add a vertical line "|" for each item. For example, the number three would be represented by three vertical lines "|||" or the number fifteen by "||||| ||||| |||||."

Tally marks are still in use today, as this is an easy way to compare two or more numerical values. The size of a number is directly proportional to its length. This notation is supported by MS Excel. Using the Excel function =REPT [text, number_to_ tally], monthly sales is presented visually as follows:

Month	Sales	Tally Representation
January	33,000	||||||||||||||||||||||||||||||||
February	35,000	|||||||||||||||||||||||||||||||||||

---

[6]   Trade Mark of Mercedes

March	35,555	|||||||||||||||||||||||||
April	36,789	|||||||||||||||||||||||||
May	38,000	|||||||||||||||||||||||||
June	39,000	|||||||||||||||||||||||||
July	40,657	|||||||||||||||||||||||||
August	41,000	|||||||||||||||||||||||||
September	42,000	|||||||||||||||||||||||||
October	42,998	|||||||||||||||||||||||||
November	43,000	|||||||||||||||||||||||||
December	50,000	|||||||||||||||||||||||||

Improving on the above representation by changing the formula to display bar chart and the rank:

To show the bar chart: =REPT("█",D5/MAX($D$5:$D$12)*25)
And for rank: =RANK(D6,$D$5:$D$14,0)

Month	Sales	Sales Representation	Rank
January	33,000	████████████████	7
February	32,000	███████████████	8
March	35,555	█████████████████	5
April	36,789	██████████████████	4
May	38,000	██████████████████	3
June	39,000	███████████████████	2
July	35,000	█████████████████	6
August	41,000	████████████████████	1

Roman numerals are another way of representing numbers. The length of the roman number, as the tally counts, also represents its numerical value. The numerical value of the number 8 is represented by the tally count "||||| |||." The same number using roman numerals is represented by "VIII."

Modern arabic numerals are irreplaceable for complex calculations, but may not be the best choice for visual representation. In fact, just by looking at the table above, it is obvious that a tally is superior to arabic representation of total sales numbers. Where a simple comparison is required, graphic representation is a better choice, but where numerical values are used for numerical calculations, arabic notation is superior.

But the story that we use to frame the numbers is also important, as studies have shown that the decisions are affected by how a story frames the number as opposed to numbers alone.

In a survey, when doctors were told that the *mortality rate* for a certain operation is six percent, they hesitated to recommend it. On the other hand, when they were told it had *survival rate* of ninety-six percent they were more inclined to recommend it to their patients.

The facts presented in both cases ware correct. Framing the facts with *survival rate* triggered more optimistic story than the phrase *mortality rate*.

War and Peace by Tolstoy Presented Visually

The highlight of Professor Edward Tufte's seminar "Presenting Data and Information" happens when he presents Charles Joseph Minard's famous statistical graph of Napoleon's disastrous Russian campaign 1812-1813, *Carte figurative des pertes successives en hommes de l'Armée Française dans la campagne de Russie 1812-1813.*

The map follows the French invasion and retreat from Russia in 1812. In *The Visual Display of Quantitative Information*, Tufte refers to Charles Joseph Minard's 1861 thematic map of Napoleon's ill-fated march on Moscow as,

> It may well be the best statistical graphic ever drawn.

He uses it as a prime example in his seminars and also in his acclaimed 1983 book, *The Visual Display of Quantitative Information*. Tufte said,

> "This is *War and Peace* as told by a visual Tolstoy."

Napoleon Bonaparte began his ill-fated 1812 invasion of the Russian Empire with 422,000 men. With each step further into Russian territory, more and more soldiers died or deserted. By the time it reached Moscow, Napoleon's army had dwindled to 100,000 men—already less than a quarter of the size it had been at the start.

During their disastrous retreat out of Russia, temperatures plunged to −37.5 °C. Nearly half the remaining survivors of the invasion were killed during the botched crossing of the Berezina River. Of the 422,000 men who set out on the invasion, barely 10,000 of them returned alive.

All this information is readily visible in the above chart, created by Minard, combining both a map of the campaign and a visual representation of the number of men remaining in Napoleon's doomed army. The thickness of the line is proportional to the number of men in the army (one millimeter equaling 10,000 men), with the gray (original color map is in beige) section representing the offensive toward Moscow, and the black line the retreat. Below, Minard also included a second chart showing the temperature on various days during the retreat.

Minard includes a description above his chart, but it is almost completely unnecessary; all the pertinent information is readily apparent from a close examination of the chart itself.

Minard was a master at the production of maps such as this that combined tremendous amounts of data with geographic representations. The most striking feature of the chart is the thinning line of soldiers, the background of the map showing the cities, and rivers the army traversed on its way into and out of Russia.

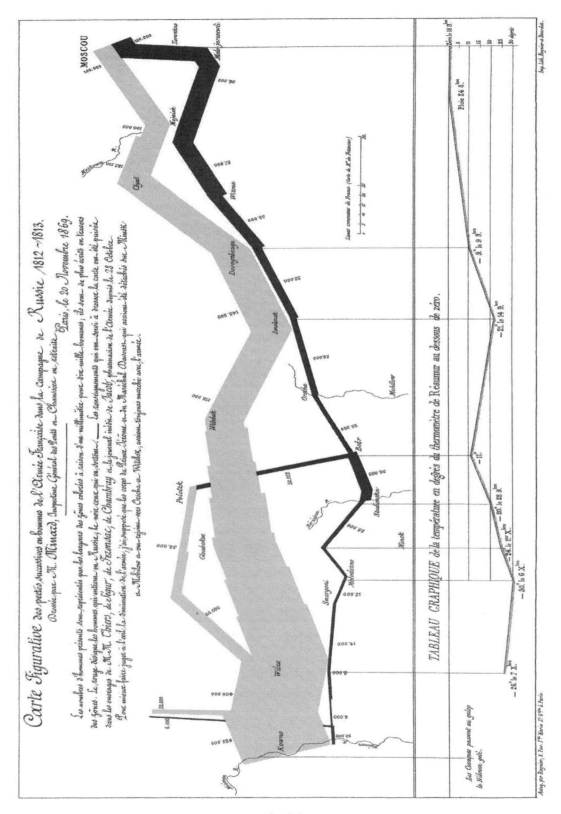

Fig 6.8

With good planning and design, this chart demonstrates how maps can operate in concert with many other types of information to create stunning displays of information.

A frustrated Napoleon had little choice but to return back to the part of Europe where he controlled for food, shelter, and supplies. Minard now traces the remnants of the *Grande Armee* as it makes its way back toward the Neiman River. In doing so, the parallel tracks of the advancing and retreating army are set next to one another, making the continuing deterioration of the army all the more visible and heart wrenching. As the army slowly made its way across barren earth (the Russians had burned food along this path while blocking other escape paths), one of the worst winters in recent memory set in.

Minard tracks the plummeting temperature against this trek on a horizontal axis at the bottom of the page, even more profoundly capturing the dire straits that the retreating army found itself in. Not surprisingly, the pitiful band of troops that returned from Russia marked the onset of the collapse of Napoleon's Continental Empire.

The graph (map) displays several variables in a single two-dimensional space like:

- The army's size, location, and direction, showing where units split off and rejoined.
- The declining size of the army. The most striking example is the crossing of the Berezina River. The Battle of Berezina took place from November 26-29, 1812, between the retreating French army of Napoleon and the Russian armies under Mikhail Kutuzov. The French suffered very heavy losses, but managed to avoid the trap and cross the river. Since then, *Bérézina* has been used in French as a synonym for disaster.
- The mild weather causing rivers to overflow its banks immediately followed by lowest temperatures during the retreat and its effect on retreating army is obvious from the map. The map is an example how good data, when treated with respect, like a good writing, convey beautifully and clearly the meaning, thus proving Tufte's assertion that data, when presented respectfully and elegantly, is not confusing, but clarifying, and it does not cause information overload.

Seeking Relationships

Relating data, or finding accurate, convenient, and useful representations of data, first involves determining the nature and structure of representation (e.g., linear regression), and then deciding how to quantify and compare the two (actual data with regression model) different representations (e.g., sum of squared errors). Once the model has been built and validated, it can be used to make predictions. Along with prediction, there should be same indication of the confidence of the prediction.

The most popular model for making prediction is multiple linear regression model. This model is used to fit a linear relationship between a quantitative dependent variable Y, also known as outcome, and a set of predictors X_1, X_2, . . . X_p, referred as independent input variables or regressors.

The basic assumption is that the following relation holds:

$$Y = \alpha + aX_1 + bX_2 + \ldots + pX_n + \beta$$

Example of Linear Regression with one variable: y = a + bx

Predictor variable = *x* (i.e., income), Response variable = *y* (spending)

Score is defined as the sum of squared errors

Need to find a and b such that *y = a+bx*

Assuming a set of numbers

Y	X
1	3
8	9
11	11
4	5
3	2
12	
15	

Scatter plot is created first to identify whether any relationship exists between the two variables. The two variables are plotted on the x- and y-axes. Each point is displayed on the scatter plot is a single observation. The scatter plot allows seeing the type of relationship that may exist between two variables.

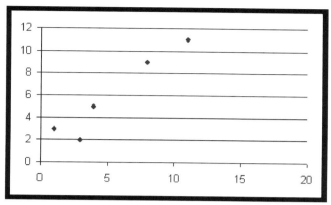

Fig. 6.9 Scatter plot

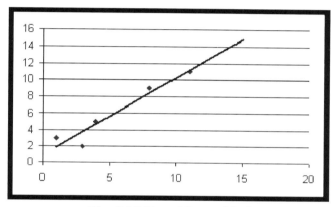

Fig. 6.10 Trend line added

Values of *a* and *b* (regression coefficients) can be calculated or derived from the trend line. The second method—deriving values for *a* and *b* from the trend line—is used here. The point at which the trend line intercepts with the y-axis is a value of *a*. The slope of the line is the value for *b* (*b* = y/x).

The derived formula for the optimal regression line (relationship between the income and spending), for the set on numbers in the table above is:

$$y = 0.8 + 1.4x$$

Comparing response values with the predicted value, it is obvious that there are differences. In the case of perfect prediction model, all the actual points would lie on the prediction line (trend line. More practical assumption is a good model, which have the actual points close to the prediction line. Confidence of the predictive accuracy is of a good model is described below:

The error or residual is defined as the difference between the actual response values and predicted values. Correlation coefficient is calculated based on the actual and predicted value. The resulting value of the correlation coefficient is between -1 and

+1, where strong positive correlations are signified with values closer to +1; strong negative relationships are closer to -1; values of close to zero indicate an absence of any relationship. When the values are squared the resulting range will be between 0 and 1. Model accuracy improves as correlation coefficient or squared correlation coefficient values get closer to 1 (1 corresponds to 100 percent accuracy).

Indeed, the notion of prototyping is closely related to validating user requirements. In software, nothing is visible until it is done, meaning that a change of mind by users (typically nonprogrammers) may be too late. Software products are typically easy to use only if they are designed by programmers for other programmers, and not for normal human beings. Or conversely, and in accordance with Grudin's law:

> *If the person responsible for using the system does not benefit, the system is doomed to failure.*

> —Jonathan Grudin, professor, Information & Computer Science Department, University of California, Irvine

Typically, users cannot design and program their own BI solutions, and are dependent on their IT colleagues. This may change in the future, as major MBA program around the world are introducing art and design in their curriculum.

Bridging the Grudin's law necessitates, for now, strong collaboration between the two groups (designers and users), from the start of the program the final solution. Designers must continuously validate that they are building the right product, as it is always cheaper to build it right the first time than to build it over and over again, especially in this case, when providing users with cognitive objects. This necessitates a human-centered design process, implemented with rapid prototyping and an iterative process to solve such a complex problem.

The concept behind a prototype is that users cannot tell what it is they want, but they can tell quickly what they don't want when they see it. A user-centered design process requires going to the places where the domain experts (users) work with the purpose of understanding people and being able to extract innovative ideas. It involves looking for the latent need, a need that has not been expressed in any way.

On the pragmatic level, it means sitting down with the users, watching them use the reports, looking out for when they are in trouble or even frustrated, looking for the moments when they are having a good time, and amplifying the good experience in the cognitive objects built for the users.

Excel is an amazing tool that could be used to build the rapid prototype. It is the best tool for executive dashboard prototyping because of its flexibility and low development costs. Creating a fully functional prototype dashboard does not require a development programmer to do, and it should not be done by the development team, as this is a QA tool used for continues requirement validation and prone to frequent changes. It should be kept out of the rigorous development version control process. The tool is used for quick modeling and user feedback should be available in a matter of days. Every time you need a dashboard project, think of a prototype in Excel.

Excel is a very powerful visualization tool, tool that can be effectively used with no or very little VBA programming. Below are simple visualization graphs:

	Jan	Feb	Mar	Apr	May	Jun	Jul	Aug	Sep	Oct	Nov	Dec
Target	30,029	31,187	28,542	34,006	36,465	35,001	39,270	39,543	33,810	39,005	39,000	39,000
Revenue	30,666	31,685	29,342	33,773	37,376	37,143	40,184	40,381	35,600	38,170	37,880	41,770
Variance to Target	$637	$498	$800	($233)	$911	$2,142	$914	$838	$1,790	($834)	($1,120)	$2,770

	2009	2010	2011
J	88	135	199
F	62	110	190
M	65	109	186
A	59	99	171
M	99	140	198
J	80	115	190
J	88	130	170
A	90	149	169
S	115	150	159
O	119	140	166
N	109	135	162
D	101	119	171

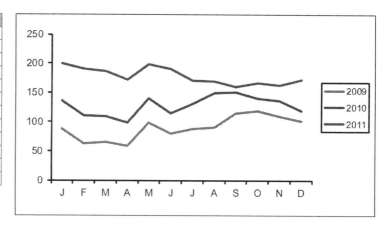

Trending graph

	Jan	Feb	Mar	Apr	May	Jun	Jul	Aug	Sep	Oct
Target	31,021	32,187	29,542	33,006	37,465	35,089	38,270	37,543	35,810	37,465
Revenue	30,685	31,685	30,342	32,773	35,376	36,143	39,184	37,381	36,600	37,170

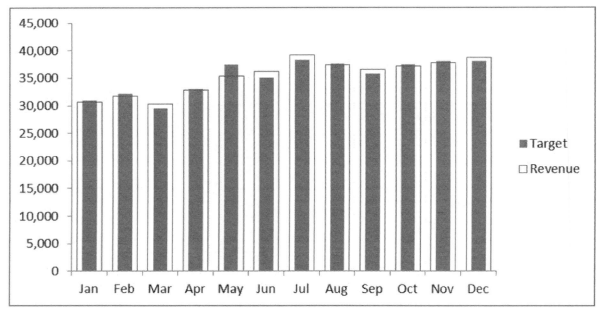

Revenue vs. Target

Many Excel users rely on Excel's default charts types and settings to produce their charts and graphs without regard for data visualization principles. Default Excel charts are not effective data visualization tools. By combining data visualization principles and advanced Excel charting techniques, Excel charts can be made into powerful data visualization tools. An Excel prototype can be used for rapid and iterative development. It is easy to create multiple alternative user interfaces, get feedback from users, or find design flaws. The prototype is used to continuously validate user requirements. The final version of the prototype is used for creating specifications for the IT department to construct the WD application.

In designing the dashboard, the goal is to craft a visual story that is in harmony with all charts and graphs and resonate seamlessly with the user's thinking. Dashboards can tell a much richer story than a simple chart. If we think of one chart as a short story, then a dashboard is a novel where characters are fully developed.

Introduction to Data Warehouse Testing

Data warehouse testing is on the rise, and qualified testers are in demand. The reasons are clear. Prominent among them is the increase in business mergers, data center migrations, compliance regulations, and management's increased focus on data and data-driven decision makings related to business intelligence (BI) initiatives. Among data warehouse testing focus is the ETL process, BI engines, and applications that rely on data warehouses.

Data-driven decisions are proving to be accurate—and they must be. In this context, testing data warehouse implementations have become of utmost significance. Organizational decisions highly depend on the enterprise data in data warehouses and must be of utmost quality. Complex business rules and transformation logic are built using ETL logic, demanding diligent and thorough testing. This book addresses challenges for DWH testing like voluminous data, heterogeneous sources, temporal inconsistency, and estimation challenges.

Preparing for the data warehouse testing process

A good understanding of data modeling and source to target data mappings help equip the QA analyst with information to develop an appropriate testing strategy. Hence, it's important that during the project's requirement analysis phase, the QA team works to understand the data warehouse implementation to the greatest extent. Data warehouse testing strategies, in most cases, will be a consortium of several smaller strategies. This is due to the nature of data warehouse implementations.

Different stages of the data warehouse implementation (source data profiling, data warehouse design, ETL development, data loading and transformations, etc.), require the testing team's participation and support. Unlike some traditional testing, test execution does not start at the end of the data warehouse implementation. In short, test execution itself has multiple phases and is staggered throughout the life cycle of the data warehouse implementation.

A key element contributing to the success of the data warehouse solution is the ability of the test team to plan, design, and execute a set of effective tests that will help identify multiple issues related to data inconsistency, data quality, data security, failures in the extract, transform and load (ETL) process, performance-related issues, accuracy of business flows and fitness for use from an end-user perspective.

Overall, the primary focus of testing should be on the end-to-end ETL process. This includes validating the loading of all required rows, correct execution of all transformations, and successful completion of the cleansing operation. The team also needs to thoroughly test SQL queries, stored procedures, or queries that produce aggregate or summary tables. Keeping in tune with emerging trends, it is also important for test team to design and execute a set of tests that are customer experience-centric.

While basic QA philosophies hold true, it's important for test teams to understand that testing a data warehouse is different from most other software testing. Since a data warehouse primary deals with data, a major portion of the test effort is spent on planning, designing, and executing tests that are data-oriented. Such tests include SQL queries, validating that ETL sessions execute as expected, that ETL tool and store procedure exceptions are handled effectively, application performance meets the SLAs, and finally, ensuring that data integration points are working as expected.

Planning and designing most of the test cases require the test team to have experience in SQL and performance testing. It will also be helpful when the team members have experience in debugging performance bottlenecks.

Data warehouse testing goals and related verification methods

Primary goals for verification over all data warehouse project testing phases include:

- *Data completeness.* Ensure that all expected data is loaded by means of each ETL procedure.
- *Data transformations.* Ensure that all data to be transformed is completed correctly according to business rules and design specifications.
- *Data quality.* Ensure that the ETL process correctly rejects, substitutes default values, corrects, ignores, and reports invalid data.
- *Performance and scalability.* Ensure that data loads and queries perform within expected time frames and that the technical architecture is scalable.
- *Integration testing.* Ensure that the ETL process functions well with other upstream and downstream processes.
- *User-acceptance testing.* Ensure the data warehousing solution meets users' current expectations and anticipates their future expectations.
- *Regression testing.* Ensure existing functionality remains intact each time a new release of ETL code and data is completed.

Listed below are a few of the many reasons to thoroughly test the data warehouse, and use a QA process that is specific to data and ETL testing:

- Source data is often huge in volume and from varied types of data repositories.
- The quality of source data cannot be assumed, and should be profiled and cleaned.
- Inconsistent and redundancy may exist in source data.
- Many source data records may be rejected; ETL/stored procedure logs will contain messages that must be acted upon.
- Source field values may be missing where they should always be present.
- Source data history, business rules, and audits of source data may not be available.
- Enterprise-wide data knowledge and business rules may not be available to verify data.
- Since data ETLs must often pass through multiple phases before loading into the data warehouse, extraction, transformation, and loading components must be thoroughly tested to ensure that the variety of data behaves as expected within each phase.
- Heterogeneous sources of data (e.g., mainframe, spreadsheets, Unix files) will be updated asynchronously through time then incrementally loaded.
- Transaction-level traceability will be difficult to attain in a data warehouse.
- The data warehouse will be a strategic enterprise resource and heavily replied upon.

Testing Phases To Be Considered for the Data Warehouse Test Strategy

Figure 3.3 shows a representative data warehouse implementation from identification of source data (lower left) to report and portal reporting (upper left). In between, several typical phases of the end-to-end data warehouse development process are depicted such as source extract to staging, dimension data to the operational data store (ODS), fact data to the data warehouse, and report and portal functions extracting data for display and reporting. The graphic illustrates that all data load programs and resulting data loads should be verified throughout the end-to-end QA process.

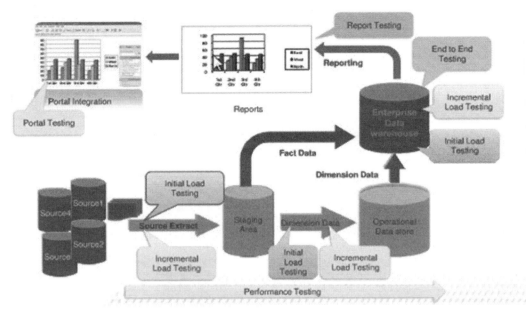

Figure 3.3 End-to-End Data Warehouse Process and Associated Testing

An end to end data warehouse test strategy is important for documenting the approach to test the warehouse implementation process. The strategy typically contains a high-level understanding of what the eventual testing workflow will be. The strategy will be used to verify and ensure that the data warehouse system meets its design specifications and other requirements.

QA skills that are helpful for data warehouse testing

- Understanding fundamental concepts of data warehousing and its place in an information management environment.
- Knowing the role of the testing process as part of data warehouse development.
- Development of data warehouse test strategies, test plans, and test cases—what they are and how to develop them, specifically for data warehouses and decision-support systems.
- Creating effective test cases and scenarios based on business and user requirements for the data warehouse.
- Participating in reviews of the data models, data mapping documents, ETL design, and ETL coding; provide feedback to designers and developers.
- Participating in the change management process and documenting relevant changes to decision support requirements.

This book aims to share the results of lessons learned during QA on many data warehouse implementation projects. For example,

Formal QA data track verifications should begin early in the ETL design and data load process, and continue through deployment and into production.

Given early access to the ETL development environment, testers can assess the quality of early data loads and offer valuable feedback to development teams. Such early access can dramatically aid preparations for formal testing and identify issues early.

Where projects utilize offshore or contract test teams, they may discover the need for more adequate and representative samples of data (production data, if possible) for test planning and test case design.

For all project stakeholders, data models, database design documents (LLDs), ETL design and data source to target mapping documents need to be kept in sync until transition.

Data warehouse test automation (particularly for regression testing) and associated tools are keys to success to support the increasingly important new life cycle models such as agile and iterative.

Planning for Data Warehouse Testing

Writing a test plan is a key to the entire data warehouse testing effort. The plan will help test engineers to validate and verify data requirements from end to end (source to target data warehouse). A primary purpose of a formal test program is to verify data requirements as stated in the:

- Business requirements document
- Data models for source and target schemas
- Source-to-target mappings
- ETL design documents

As requirement documents specifications are the *what* for ETL development, the test plan can serve as the *what* for the test process. The test plan describes what QA staff will develop to verify that the data warehouse meets requirements. Properly constructed, the test plan is a contract between the QA team and all other project stakeholders.

In addition to the data requirements, we list below further considerations for the test planning effort.

- Configuration management system
- Project schedule
- Data quality verification process
- Incident and error handling system
- QA staff resources estimates and training needs
- Testing environment budget and plan
- Test tools
- Test objectives
- QA roles and responsibilities
- Test deliverables
- Test tasks
- Defect reporting requirements
- Entrance criteria that should be met before formal testing commences
- Exit criteria that should be met before formal testing is completed

Planning Tests for Common Data Warehouse Issues

Following are some of the issues/defects that may be found during data warehouse testing. Planning in advance on how to identify issues during tests is important.

- Inadequate ETL and stored procedure design documentation to aid in test planning.
- Field values are null when specified as *Not Null*.
- Field constraints and SQL not coded correctly for the ETL tool.
- Excessive ETL errors discovered after entry to formal QA.
- Source data does not meet table mapping specifications (ex. dirty data).
- Source-to-target mappings: (1) often not reviewed before implementation, (2) are in error or (3) not consistently maintained throughout the development life cycle.
- Data models are not adequately maintained during the development life cycle.
- Duplicate field values are found in either source or target data when defined in mapping specifications to be *distinct*.
- ETL SQL/transformation errors leading to missing rows and invalid field values.
- Constraint violations exist in source (perhaps could be found through data profiling).
- Target data is incorrectly stored in nonstandard formats.
- Primary or foreign key values are incorrect for important relationship linkages.

Source-to-Target Data Mapping Explained

The data movement discipline focuses on the movement of data between systems. In this context, *systems* include external data sources, operational systems, and analytic data stores. Data movement encompasses the extract-transform-load (ETL) facilities used for bulk data movement. It also includes mechanisms to support continuous movement of discrete records, rows, or messages between systems. What are mapping and data movement best practices?

Data movement ETL best practices provide a guide for the analysis, design, and development of data movement processes that are consistent, usable, and high quality.

Common data mapping best practices are important for the following reasons:

- They lead to greater consistency, which subsequently leads to greater productivity;
- They reduce ongoing maintenance costs; and
- They improve readability of software, making it easier for developers to understand new code more quickly 1987

Some data mapping and data movement best practice goals:

- Introduce common, consistent data movement analysis, design, and coding patterns,
- Develop reusable, enterprise-wide analysis, design, and construction components through data movement modeling processes using data movement tools, to ensure an acceptable level of data quality per business specifications,
- Introduce best practices and consistency in coding and naming standards.
- Reduce costs to develop and maintain analysis, design and source code deliverables, and
- Integrate controls into the data movement process to ensure data quality and integrity.
- An ETL conceptual data movement model should be created as part of the information management strategy. This model is part of the business model and shows what data flows into, within, and out of the organization.

This is a *high-level* model that shows movement of data from one application to another. Order entry systems may capture customer name, address information, and later send it to the application, which handles billing and accounting, so that bills can be sent out and money collected.

Think of the conceptual data model as an architect's conceptual drawing of a house. It provides a good idea of what is required, with very little additional detail.

- A logical data movement model should show data movement requirements at the dataset (entity/table)-level. It should detail the transformation rules and target logical datasets (entity/tables). This model is still considered technology independent.

It should show each interface between systems. An order-entry system—which sends customer names and address information to the billing application—may actually send three or four files. Each file is called an interface. Each interface should be shown in the logical data movement model.

The focus at the logical level is on the capture of actual source tables and proposed target stores.

The logical data movement model should be supported by a source to target mapping document, which should include business rules and data transformation rules. Below, we display such a mapping document.

Target Table	Target Column	Data-type	Source Table	Source Column	Data-type	Expression	Default Value If Null	Data Issues/Quality/Comments
Course	Uniqueid	Varchar2(20)	Course	Uniqueid	Varchar2(20)		Not NULL	PK
Course	Subject Area Admin Uniqueid	Varchar2(20)	Course	Subject Area	Varchar2(20)			
Course	Academic Specialization	Varchar2(20)	Course (admin) or Subject Area Administration (admin2)	Academic Specialization	Varchar2(20)	DECODE(COURSE_ACAD_SPN,NULL,SUBJ_AREA_ACAD_SPN,COURSE_ACAD_SPN)		If the academic specialization from course is null, use academic specialization from subject area administration
Course	Special Fee	Varchar2(20)	Course	Special Fee	Varchar2(20)	decode(SPECIAL_FEE_UNIQUEID_in,null,NO_SPECIAL_FEE',SPECIAL_FEE_UNIQUEID_in)		If the special fee is null, set = 'NO_SPECIAL_FEE'
Course	Department	Varchar2(20)	Course (admin) or Acad Initiative Sbj Area (admin)	Department	Varchar2(20)	DECODE(COURSE_DEPARTMENT_UNIQUEID,null,ACAD_DEPARTMENT_UNIQUEID,COURSE_DEPARTMENT_UNIQUEID)		If the department is null, use deparment from ACAD_INITIATIVE_SBJ_AREA
Course	Current Flag	Varchar2(3)	Course	Compare sysdate/First Effective Date/Last Effective Date	Date	decode(DATE_COMPARE(SYSDATE, FIRST_EFFECTIVE_DATE),1,decode(DATE_COMPARE(LAST_EFFECTIVE_DATE, sysdate),0,'YES', 1,'YES', 1,'NO',null,'YES'),0,decode(DATE_COMPARE(LAST_EFFECTIVE_DATE, sysdate),0,'YES', 1,'YES',-1,'NO',null,'YES')),-1,'NO')		

Topics for the Data Warehouse Test Plan

The Test plan should list the scope of each test category (ex. unit, system, integration testing), including what is the entrance criteria, how test cases will be prepared, what test data will be used, what test scripts are required, who will execute test scripts, how defects will be managed, how test results will be managed, and what is the test exit criteria.

Following are topics that should be considered for the data warehouse test plan:

- *Test objectives* should summarize data verification objectives, testing scope, and staff roles and responsibilities.
- *Test strategy* should refer to the data requirements acceptance criteria that were developed as part of the requirements definition phase, and the test environments that will be used.
- *Test data strategy* should list the overall approach for creating test data. Try not to use large volumes of data for unit, system, integration, regression, and quality assurance testing. Most projects can adequately complete data warehouse tests with a subset of data. Failure to do this will add significantly to the test period and will not likely have an additional benefit. The extra time it takes to identify a complete set of test data early in the project will be paid back many times by reducing the time to back-up/restore data during testing.
- *Test deliverables* should specify what will be produced during testing.
- *Resource plan* should list all project roles and responsibilities, level of effort, and other test resource requirements.
- *Training considerations* should list any training that might be required so the test team can complete testing.
- *Test schedule* should identify a test schedule that clearly defines when all testing is expected to occur. This schedule may be included in the project schedule.

Those involved in test planning should consider the following verifications as primary among those planned for various phases of the data warehouse loading project.

- Verify data mappings, source to target
- Verify that all tables and specified fields were loaded from source to staging
- Verify that primary and foreign keys were properly generated using sequence generator or similar
- Verify that not-null fields were populated
- Verify no data truncation in each field
- Verify data types and formats are as specified in design phase
- Verify no unexpected duplicate records in target tables.

- Verify transformations based on data table low level design (LLDs—usually text documents describing design direction and specifications)
- Verify that numeric fields are populated with correct precision
- Verify that each ETL session completed with only planned exceptions
- Verify all cleansing, transformation, error and exception handling
- Verify stored procedure calculations and data mappings

Common QA Tasks for the Data Warehouse Team

During the data warehouse testing life cycle, many of the following tasks may be typically be executed by the QA team. It is important to plan for those tasks below that are keys to the project's success.

- Complete test data acquisition and baseline all test data.
- Create test environments.
- Document test cases.
- Create and validate test scripts.
- Conduct unit testing and confirm that each component is functioning correctly.
- Conduct testing to confirm that each group of components meet specification.
- Conduct quality assurance testing to confirm that the solution meets requirements.
- Perform load testing, or performance testing, to confirm that the system is operating correctly and can handle the required data volumes and that data can be loaded in the available load window.
- Specify and conduct reconciliation tests to manually confirm the validity of data.
- Conduct testing to ensure that the new software does not cause problems with existing software.
- Conduct user acceptance testing to ensure that business intelligence reports work as intended.
- Carefully manage scope to ensure that perceived defects are actually requirement defects and not something that would be "nice to have, but we forgot to ask."
- Conduct a release test and production readiness test.
- Ensure that the ongoing defect management and reporting is effective.
- Manage testing to ensure that each follows testing procedures and software testing best practices.
- Establish standard business terminology and value standards for each subject area.
- Develop a business data dictionary that is owned and maintained by a series of business-side data stewards. These individuals should ensure that all terminology is kept current and that any associated rules are documented.

- Document the data in your core systems and how it relates to the standard business terminology. This will include data transformation and conversion rules.
- Establish a set of data acceptance criteria and correction methods for your standard business terminology. This should be identified by the business-side data stewards and implemented against each of your core systems (where practical).
- Implement a data profiling program as a production process. You should consider regularly measuring the data quality (and value accuracy) of the data contained within each of your core operational systems.

Data integration planning (Data model and DB low-level design)

- Gain understanding of data to be reported by the application (e.g., profiling) and the tables upon which each user report will be based upon.
- Review and understand the data model—gain understanding of keys, flows from source to target.
- Review, understand data LLDs and mappings; add, update sequences for all sources of each target table.

ETL planning and testing (source inputs and ETL design)

- Participate in ETL design reviews.
- Gain in-depth knowledge of ETL sessions, the order of execution, restraints, and transformations.
- Participate in development ETL test case reviews.
- After ETLs are run, use checklists for QA assessments of rejects, session failures, and errors.

Assess ETL logs: session, workflow, errors

- Review ETL workflow outputs, source-to-target counts.
- Verify source to target mapping docs with loaded tables using TOAD and other tools.
- After ETL runs or manual data loads, assess data in every table with focus on key fields (dirty data, incorrect formats, duplicates, etc.). Use TOAD, Excel tools. (SQL queries, filtering, etc.)

Considerations for Selecting Data Warehouse Testers

Members of the QA staff who will plan and execute data warehouse testing should have many of the following skills and experiences.

- Over five years of experience in testing and development in the fields of data warehousing, client server technologies, which includes over five years of extensive experience in data warehousing with Informatica, SSIS or other ETL tools.
- Strong experience in Informatica or SQL Server, stored procedure and SQL testing.
- Expertise in unit and integration testing of the associated ETL or stored procedure code.
- Experience in creating data verification unit and integration test plans and test cases based on technical specifications.
- Demonstrated ability to write complex multi-table SQL queries.
- Excellent skills with OLAP, ETL, and business intelligence.
- Experience with dimensional data modeling using Erwin Modeling star join schema/snowflake modeling, fact and dimensions tables, physical and logical data modeling.
- Experience in OLAP reporting tools like Business Objects, SSRS, OBIEE or Cognos.
- Expertise in data migration, data profiling, data cleansing.
- Hands on experience with source-to-target mapping in enterprise data warehouse environment. Responsible for QA tasks in all phases of the system development life cycle (SDLC), from requirements definition through implementation, on large-scale, mission critical processes; excellent understanding of business requirements development, data analysis, relational database design, systems development methodologies, business/technical liaising, workflow and quality assurance.
- Experienced in business analysis, source system data analysis, architectural reviews, data validation, data testing, resolution of data discrepancies and ETL architecture. Good knowledge of QA processes.
- Familiarity with performance tuning of targets databases and sources system.
- Extensively worked in both UNIX (AIX/HP/Sun Solaris) and Windows (Windows SQL Server) platforms.
- Good knowledge of UNIX Shell Scripting and understanding of PERL scripting.
- Experience in Oracle 10g/9i/8i, PL/SQL, SQL, TOAD, Stored Procedures, Functions and Triggers.

QA Checklists for Data Warehouse Quality Verification

This section describes testing guidelines and steps for verifying data, ETL processes, and SQL during the construction, unit testing, system and integration testing of an application's data warehouse operational tables and data mart.

An Overview of Data Warehouse Testing

A data warehouse is a repository of transaction data that has been extracted from original sources and transformed so that query, analysis, and reporting on trends within historic data are both possible and efficient. The analyses provided by data warehouses may support an organization's strategic planning, decision support, or monitoring of outcomes of chosen strategies.

Typically, data that is loaded into a data warehouse is derived from diverse sources of operational data, which may consist of data from databases, feeds, application files (such as office productivity software files) or flat files. The data must be extracted from these diverse sources, transformed to a common format, and loaded into the data warehouse.

Extraction, transformation and loading (ETL) is a critical step in any data warehouse implementation, and continues to be an area of major importance over the life of the warehouse due to recurrent warehouse updating. Once a data warehouse is populated, front-end systems must be tested to facilitate querying, analysis, and reporting. The data warehouse front-end may provide a simple presentation of data based on queries, or may support sophisticated statistical analysis options. Data warehouses may have multiple front-end applications, depending on the various profiles within the user community.

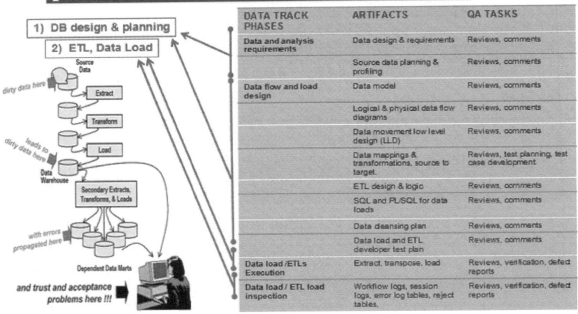

An effective data warehouse testing strategy focuses on the main structures within the data warehouse architecture:

1) The ETL layer
2) The full data warehouse
3) The front-end data warehouse applications

Each of these units must be treated separately and in combination, and since there may be multiple components in each (multiple feeds to ETL, multiple databases or data repositories that constitute the warehouse, and multiple front-end applications), these subsystems must be individually validated.

Verify and Maintain the Data Low Level Design (LLD)

A first level of testing and validation begins with the formal acceptance of the logical data model and low level design (LLD). All further testing and validation will be based on the understanding of each of the data elements in the model.

Data elements that are created through a transformation or summary process must be clearly identified and calculations for each of these data elements must be clear and easily interpreted.

During the LLD reviews and updates, special consideration should be given to typical modeling scenarios that exist in the project. Examples follow:

Verify that many-to-many attribute relationships are clarified and resolved.

- Verify the types of keys that are used—surrogate keys versus natural keys.
- Verify that the business analyst/DBA reviewed with ETL architect and developers (application) the lineage and business rules for extracting, transforming, and loading the data warehouse.
- Verify that all transformation rules, summarization rules, and matching and consolidation rules have clear specifications.
- Verify that specified transformations, business rules and cleansing specified in LLD and other application logic specs have been coded correctly in ETL, JAVA, and SQL used for data loads.
- Verify that procedures are documented to monitor and control data extraction, transformation, and loading. The procedures should describe how to handle exceptions and program failures.
- Verify that data consolidation of duplicate or merged data was properly handled.
- Verify that samplings of domain transformations will be taken to verify they are properly changed.

- Compare unique values of key fields between source data and data loaded to the warehouse. This is a useful technique that points out a variety of possible data errors without doing a full validation on all fields.
- Validate that target data types are as specified in the design and/or the data model.
- Verify how sub-class/super-class attributes are depicted.
- Verify that data field types and formats are specified.
- Verify that defaults are specified for fields where needed.
- Verify that processing for invalid field values in source are defined.
- Verify that expected ranges of field contents are specified where known.
- Verify that keys generated by the sequence generator are identified.
- Verify that slowly changing dimensions are described.

Analyze Source Data before and after Extraction to Staging

Testers should extract representative data from each source file (before or after extract to staging tables) and confirm that the data is consistent with its definition; QA can discover any anomalies in how the data is represented and write defect reports where necessary. The objective is to discover data that does not meet data quality factors as described in specifications. See list below and Table 1.

This verification process will be used for temp tables used in a step process for data transformations, cleaning, etc.

- Verify that the scope of values in each column is within specifications.
- Identify unexpected values in each field.
- Verify relationships between fields.
- Identify frequencies of values in columns and whether these frequencies make sense.

Inputs: Application source data models and low level data design, data dictionaries, data attribute sources.

Outputs: Newly discovered attributes, undefined business rules, data anomalies such as fields used for multiple purposes.

Techniques and Tools: Data extraction software, business rule discovery software, data analysis tools.

Process Description:

- Extract representative samples of data from each source or staging table.
- Parse the data for the purpose of profiling.

- Verify that not-null fields are populated as expected.
- Structure discovery—Does the data match the corresponding metadata? Do field attributes of the data match expected patterns? Does the data adhere to appropriate uniqueness and null value rules?
- Data discovery—Are the data values complete, accurate and unambiguous?
- Relationship discovery—Does the data adhere to specified required key relationships across columns and tables? Are there inferred relationships across columns, tables or databases? Is there redundant data?
- Verify that all required data from the source was extracted. Verify that extraction process did not extract more or less data source than it should have.
- Verify or write defects for exceptions and errors discovered during the ETL process.
- Verify that extraction process did not extract duplicate data from the source (usually this happens in repeatable processes where at point zero we need to extract all data from the source file, but the during the next intervals we only need to capture the modified, and new rows).
- Validate that no data truncation occurred during staging.
- Utilize a data profiling tool or methods that show the range and value distributions of fields in the source data. This is used to identify any data anomalies from source systems that may be missed even when the data movement is correct.
- Validation and Certification Method. It is sufficient to identify the requirements and count (via SQL) the number of rows that should be extracted from the source systems. The QA team will also count the number of rows in the result/target sets and match the two for validation. The QA team will maintain a set of SQL statements that are automatically run at this stage to validate that no duplicate data have been extracted from the source systems.

Table 1: Data Quality Factors

FACTOR	DESCRIPTION	EXAMPLE
Data Consistency Issues:		
Varying Data Definitions	The data type and length for a particular attribute may vary in files or tables though the semantic definition is the same.	Account number may be defined as: Number (9) in one field or table and Varchar2(11) in another table
Misuse of Integrity Constraints	When referential integrity constraints are misused, foreign key values may be left "dangling" or inadvertently deleted.	An account record is missing but dependent records are not deleted.
Nulls	Nulls when field defined as "not-null."	The company has been entered as a null value for a business. A report of all companies would not list the business.

Data Completeness Issues:

Missing data	Data elements are missing due to a lack of integrity constraints or nulls that are inadvertently not updated.	An account date of estimated arrival is null thus impacting an assessment of variances in estimated/actual account data.
Inaccessible Data	Inaccessible records due to missing or redundant identifier values.	Business numbers are used to identify a customer record. Because uniqueness was not enforced, the business ID (45656) identifies more than one customer.
Missing Integrity Constraints	Missing constraints can cause data errors due to nulls, nonuniqueness, or missing relationships.	Account records with a business identifier exist in the database but cannot be matched to an existing business.

Data Correctness Issues:

Loss Projection	Tables that are joined over non key attributes will produce non existent data that is shown to the user.	Lisa Evans works in the LA office in the Accounting department. When a report is generated, it shows her working in IT department.
Incorrect Data Values	Data that is misspelled or inaccurately recorded.	123 Maple Street is recorded with a spelling mistake and a street abbreviation (123 Maple St)
Inappropriate Use of Views	Data is updated incorrectly through views.	A view contains non key attributes from base tables. When the view is used to update the database, null values are entered into the key columns of the base tables.
Disabled Integrity Constraints	Null, non unique, or out of range data may be stored when the integrity constraints are disabled.	The primary key constraint is disabled during an import function. Data is entered into the existing data with null unique identifiers.
Nonduplication	Testing should be conducted to determine if there's duplication of data where there should not be.	Duplicate rows or column data.
Misuse of Integrity Constraints	Check whether null or foreign key constraints are inappropriate or too restrictive.	Check constraint only allows hard coded values of C, A, X, and Z. But a new code B cannot be entered.

Data Comprehension Issues:		
Data Aggregation	Aggregated data is used to represent a set of data elements.	One name field is used to store surname, first name, middle initial, and last name (e.g., John, Hanson, Mr.).
Cryptic Object Definitions	Database object (e.g., column) has a cryptic, unidentifiable name.	Customer table with a column labeled *c_avd*. There is no documentation as to what the column might contain.
Unknown or Cryptic Data	Cryptic data stored as codes, abbreviations, truncated, or with no apparent meaning.	Shipping codes used to represent various parts of the customer base (01, 02, 03). No supporting document to explain the meaning of the codes.
Accuracy	Data will be matched against business rules.	Boundary values (low, high) will be identified for relevant fields and compared with expectations.
Completeness	Data will be assessed to verify that all required is present. Missing rows will be identified; Null values will be identified in data elements where a value is expected.	
Precision	Precision testing is conducted to evaluate the level of data not sufficiently precise based on specifications.	

Verify Corrected, Cleaned, Source Data in Staging

This step works to improve the quality of existing data in source files or *defects* that meet source specs but must be corrected before load.

Inputs:

- Files or tables (staging) that require cleansing; data definition and business rule documents, data map of source files and fields; business rules, and data anomalies discovered in earlier steps of this process.
- Fixes for data defects that will result in data does not meet specifications for the application DWH.

Outputs: Defect reports, cleansed data, rejected or uncorrectable data.

Techniques and Tools: Data reengineering, transformation, and cleansing tools, MS Access, Excel filtering.

Process Description: In this step, data with missing values, known errors, and suspect data is corrected. Automated tools may be identified to best to locate, clean/correct large volumes of data.

- Document the type of data cleansing approach taken for each data type in the repository.
- Determine how *uncorrectable* or suspect data is processed, rejected, maintained for corrective action. SMEs and stakeholders should be involved in the decision.
- Review ETL defect reports to assess rejected data excluded from source files or information group targeted for the warehouse.
- Determine if data not meeting quality rules was accepted.
- Document in defect reports, records and important fields that cannot be easily corrected.
- Document records that were corrected and how corrected.
- Certification Method: Validation of data cleansing processes could be a tricky proposition, but certainly doable. All data cleansing requirements should be clearly identified. The QA team should learn all data cleansing tools available and their methods. QA should create various conditions as specified in the requirements for the data cleansing tool to support and validate its results. QA will run a volume of real data through each tool to validate accuracy as well as performance.

Verifying Matched and Consolidated Data

There are often ETL processes where data has been consolidated from various files into a single occurrence of records. The cleaned and consolidated data can be assessed to very matched and consolidated data.

Much of the ETL heavy lifting occurs in the transform step where combined data, data with quality issues, updated data, surrogate keys, and build aggregates are processed.

Inputs: Analysis of all files or databases for each entity type.

Outputs:

- Report of matched, consolidated, related data that is suspect or in error.

- List of duplicate data records or fields.
- List of duplicate data suspects.

Techniques and Tools: Data matching techniques or tools; data cleansing software with matching and merging capabilities.

Process Description:

- Establish match criteria for data. Select attributes to become the basis for possible duplicate occurrences (e.g., names, account numbers).
- Determine the impact of incorrectly consolidated records. If the negative impact of consolidating two different occurrences such as different customers into a single customer record exists, submit defect reports. The fix should be higher controls to help avoid such consolidations in the future.
- Determine the matching techniques to be used. Exact character match in two corresponding fields such as wild card match, key words, close match, etc.
- Compare match criteria for specific record with all other records within a given file to look for intra-file duplicate records.
- Compare match criteria for a specific record with all records in another file to seek inter-file duplicate records.
- Evaluate potential matched occurrences to assure they are, in fact, duplicate.
- Verify that consolidated data into single occurrences is correct.
- Examine and rerelate data related to old records being consolidated to new occurrence-of-reference record. Validate that no related data was overlooked.

Verify Transformed/Enhanced/Calculated Data to Target Tables

At this stage, base data is being prepared for loading into the application operational tables and the data mart. This includes converting and formatting cleansed, consolidated data into the new data architecture, and possibly enhancing internal operational data with external data licensed from service providers.

The objective is to successfully map the cleaned, corrected and consolidated data into the DWH environment.

Inputs: Cleansed, consolidated data; external data from service providers; business rules governing the source data; business rules governing the target DWH data; transformation rules governing the transformation process; DWH or target data architecture; data map of source data to standardized data.

Output: Transformed, calculated, enhanced data; updated data map of source data to standardized data; data map of source data to target data architecture.

Techniques and Tools: Data transformation software; external or online or public databases.

Process Description:

- Verify that the data warehouse construction team is using the data map of source data to the DWH standardized data, verify the mapping.
- Verify that the data transformation rules and routines are correct.
- Verify the data transformations to the DWH and assure that the processes were performed according to specifications.
- Verify that data loaded in the operational tables and data mart meets the definition of the data architecture including data types, formats, accuracy, etc.
- Develop scenarios to be covered in Load Integration Testing.
- Count Validation: Record Count Verification DWH backend/Reporting queries against source and target as an initial check.
- Dimensional Analysis: Data integrity exists between the various source tables and parent/child relationships.
- Statistical Analysis: Validation for various calculations.
- Data Quality Validation: Check for missing data, negatives and consistency. Field-by-field data verification will be done to check the consistency of source and target data.
- Granularity: Validate at the lowest granular level possible (lowest in the hierarchy, e.g., Country-City-Sector—start with test cases).
- Dynamic Transformation Rules and Tables: Such methods need to be checked continuously to ensure the correct transformation routines are executed. Verify that dynamic mapping tables and dynamic mapping rules provide an easy, documented, and automated way for transforming values from one or more sources into a standard value presented in the DWH.
- Verification Method: The QA team will identify the detailed requirements as they relate to transformation and validate the dynamic transformation rules and tables against DWH records. Utilizing SQL and related tools, the team will identify unique values in source data files that are subject to transformation. The QA team identifies the results from the transformation process and validate that such transformation have accurately taken place.

Front-end UI and Report Testing Using Operational Tables and Data Mart

<u>End-user reporting</u> is a major component of the application project. The report code may run aggregate SQL queries against the data stored in the data mart and/or the operational tables then display results in a suitable format either in a Web browser or on a client application interface.

Once the initial view is rendered, the reporting tool interface provides various ways of manipulating the information such as sorting, pivoting, computing subtotals, and adding view filters to slice-and-dice the information further. Special considerations such as those below will be prepared while testing the reports:

The ETL process should be complete; the data mart must be populated and data quality testing should be largely completed.

The front-end will use a SQL engine which will generate the SQL based on the how the dimension and fact tables are mapped. Additionally, there may be global or report-specific parameters set to handle very large database (VLDB)-related optimization requirements. As such, testing of the front-end will concentrate on validating the SQL generated; this, in turn, validates the dimensional model and the report specification vis-à-vis the design.

Unit testing of the reports will be conducted to verify the layout format per the design mockup, style sheets, prompts and filters, attributes, and metrics on the report.

Unit testing will be executed both in the desktop and Web environment.

System testing of the reports will concentrate on various report manipulation techniques like the drilling, sorting, and export functions of the reports in the Web environment.

Reports and/or documents need special consideration for testing because they are high visibility reports used by the top analysts and because they have various charts, gauges and data points to provide a visual insight to the performance of the organization in question.

There may be some trending reports, or more specifically called comp reports, that compare the performance of an organizational unit over multiple time periods. Testing these reports needs special consideration especially if a fiscal calendar is used, instead of an English calendar for time period comparison.

For reports containing derived metrics, special focus should be paid to any subtotals. The subtotal row should use a *smart-total*, i.e., do the aggregation first and then do the division instead of adding up the individual cost per click of each row in the report.

Reports with *nonaggregate-able* metrics (e.g., inventory at hand) also need special attention to the subtotal row. For example, it should not add up the inventory for each week and show the inventory of the month.

During unit testing, all data formats should be verified against a standard or rules. For example, metrics with monetary value should show the proper currency symbol, decimal point precision (at least two places), and the appropriate positive or negative symbol. For example, negative numbers should be shown in red and enclosed in braces.

During system testing, while testing the drill-down capability of reports, care will be taken to verify that the subtotal at the drill-down report matches with the corresponding row of the summary report. At times, it is desirable to carry the parent attribute to the drill-down report; verify the requirements for this.

When testing reports containing conditional metrics, care will be taken to check for *outer join condition*, i.e., nonexistence of one condition is reflected appropriately with the existence of the other condition.

Reports with multilevel sorting will get special attention for testing especially if the multilevel sorting includes both attributes and metrics to be sorted.

Reports containing metrics at different dimensionality and with percent-to-total metrics and/or cumulative metrics needs will get special attention to check that the subtotals are hierarchy-aware (i.e., they "break" or "reinitialized" at the appropriate levels).

Data displays on the business views and dashboard are as expected.

Users can see reports according to their user profile authentications and authorizations.

Where graphs and data in tabular form exist, both should reflect consistent data.

Operational Table and Data Mart: A Build Sanity Test

1. *Session Completions:* All workflow sessions completed successfully using the log viewer.
2. *Source-to-Target Counts:* This process verifies that the number of records in the source system matches the number of records received and ultimately processed into the data warehouse. If lookups are involved in the ETL process, the count between source and target will not match. The ETL session log and target table counts are compared.
3. *Source-to-Target Data Verification*: The process verifies that all source and reference tables have data before running ETLs. We verify that all target tables were truncated before the load unless target tables are updated. This process verifies that the source field threshold is not subject to truncation during the transformation or loading of data.

4. *Field-to-Field Verification:* This process verifies the field values from the source system to target. This process ensures that the data mapping from the source system to the target is correct, and that data sent has been loaded accurately.

5. *ETL Exception Processing:* Exception processing verification looks for serious data errors that would cause system processing failures or data corruption. An exception report verifying the number and types of errors encountered is produced and reviewed for additional processing and/or reporting to the customer.

There are two primary types of exception process:

1. Database Exception

    *   Not Null—source column is null while target is not null.
    *   Reference Key—the records coming from the source data do not have a corresponding parent key in the parent table.
    *   Unique Key—the record already exists in the target table.
    *   Check Constraint—these constraints enforce domain integrity by limiting the values that are accepted by a column

2. Business Rule Exception

    These are the exceptions thrown based on certain business rules defined for specific data elements or group of data elements.

    *   ETL process utilizes a single exception table to capture the exceptions from various ETL sessions and an error lookup table which has various error codes and their description.
    *   We check the exception process using the session log and exception table.

Sanity Test: Exit and Suspension Criteria

*   No critical defects unfixed; no more than 3 high severity defects.
*   80 percent or more of build functionality can be tested—functionality might fail because of Java/report code.
*   Platform performance is such that test team can productively work to schedule.
*   Fewer than 15 percent of build fixes failed.

Useful Queries to Verify Source to Target Data Loads

After data warehouse tables are built and populated, they are ready for verification. How do you initially test a data warehouse? Where do you start? There are often a few billion numbers and strings in the data warehouse; how do you verify them all?

Can you just verify a few rows and then declare that the data warehouse is fit for purpose?

No.

So how do we do it?

The difficult task is to find where the inevitable errors are. A primary method to test is by looking at grand totals for each table's individual fields (where numerical are possible). Let's say you have this fact table:

dim1_key	dim2_key	measure1	measure2
1	1	78.45	3.70
1	2	350.00	4.05
2	1	203.98	2.15
2	2	19.80	0.40

Figure 1. A fact table with two measures and two dim key columns

Step 1. You ignore the dim key column and just get the sum of measure1 and the sum of measure2 separately. This can be done where the load from source to target represented no changes to field values and no records should be rejected.

Say the total is 5,000,000 for measure1 and 4,000 for measure2. You then verify these 2 numbers against the source system. If you don't see the same results, don't move to the next step. If the grand total is wrong, there's no point getting the break down. Of course, you must take into consideration any business rules that were applied when

Step 2. You then check the breakdown of the dimension attribute. Let's say your dim1 is like this:

dim1_key	attribute1	attribute2	attribute3
1	Value1a	Value2a	Value3a
2	Value1b	Value2b	Value3b
3	Value1c	Value2c	Value3c

Figure 2. A dim table with 1 dim key column and 3 attribute columns

You do something like this:

```
select d.attribute1, sum(f.measure1), sum(f.measure2)
select d.attribute1, sum(f.measure1), sum(f.measure2)
from fact1 f
inner join dim1 d on d.dim1key = f.dim1key
group by d.attribute1
```

Say the output is like this:

attribute1	measure1	measure2
Value1a	14,038.69	581.77
Value1b	7,535.01	319.04
Value1c	3,021.87	140.55
Total	24,595.57	1,041.36

Figure 3. Sum of measure1 and measure2 by attribute1

Of course the SQL doesn't give you the total; you calculate the total yourself. You then verify this output against the source system. (a) Is there a missing record, or an extra record? (b) Are the numbers for measure1 and measure2 all correct?

Step 2 is actually quite a big task. The number of queries is the same as the number of attributes/fields in the table. For a data warehouse with ten dimensions containing ten attributes each, you will have to produce one hundred queries. You may be able to simplify this process over time, but these are the basics.

Step 3. You verify the bottom levels (the leaf level). Obviously, we can't verify all rows in the warehouse because of the volume. A fact table can contain a billion rows.

So here we are forced to take a sample. So get a set of attribute value, i.e., dim1. attribute1 = value1, dim2.attribute2 = value2, etc. When you apply this filter to the fact table, you get a smaller number of rows, like this:

```
select f.measure1, f.measure2
from fact1
join dim1 d1 on d1.dim1key = f.dim1key
join dim2 d2 on d2.dim2key = f.dim2key
where d1.attribute1 = 'value1a' and d2.attribute2 = 'value2a'
```

So instead of one billion rows, we only extract something like one thousand rows for the test. So from this (value1a, value2a) set, which contains one thousand rows, we chose three rows. We then verify all the measures in these three rows. Then we do the same thing with fact2, fact3, etc., until we have done all the fact tables in the data warehouse.

Finding the error rows

When completing Step 1 for example, you found that the total doesn't match with the source systems. So how do you find out the incorrect rows? One way is using a cube. You can use reports, but it would be a tedious process. Whereas using a cube is very easy. Following is a process for doing so.

It's likely that many of us have been involved in a data warehouse project where they never tested the totals per field. The System Integration Test only tested one or a few rows. And the User Acceptance Test only tested one or a few rows, too. While those few rows might be correct, other rows can be wrong. So it is extremely important to test the total for all measures.

It is not difficult to get the totals for each measure. Finding the error rows is the difficult task. But with a cube, imagine if you have the total of the error in measure1 is $18,755. This is the total difference between the source systems and the data warehouse for measure1. You can then drill down by any dimension/attribute you like. Say you drill down by cost center. You will find something like this:

cost centre	measure1
cost centre 1	0.00
cost centre 2	0.00
cost centre 3	0.00
cost centre 4	6,700.00
cost centre 5	0.00
cost centre 6	12,055.00
cost centre 7	0.00
cost centre 8	0.00

Figure 4. Drill down of measure1 discrepancy by cost center.

In figure 4, we are clear that the cause of this $18,755 difference is cost centers 4 and 6. All the other cost centers are spot on. They fully reconcile, i.e., the source is equal to the target. In cost centers 4 and 6 that the total are not the same.

You can then drill down on cost center 4 by another attribute, say, country, like this:

country	measure1
country 1	0.00
country 2	0.00
country 3	0.00
country 4	0.00
country 5	5,000.00
country 6	0.00
country 7	0.00
country 8	2,700.00

Figure 5. Drill down of measure1 discrepancy by country

So you know that the $6700 difference in cost center 4 is caused by country 5 and country 8.

If you keep drilling down, eventually you will arrive at leaf level, the most detailed level (say, it's transaction level), which is like this:

transaction	measure1
transaction 1	0.00
transaction 2	0.00
transaction 3	0.00
transaction 4	0.00
transaction 5	40.00
transaction 6	0.00
transaction 7	0.00
transaction 8	0.00

Figure 6. Detail/transaction level

Once we find which transactions are causing the difference, we can investigate the transaction in the source system and compare it to the data warehouse. Is it because of currency conversion? Is it because of business rules? Is the because of incorrect business/surrogate key?

Once we found the cause of the issue, we can correct the data integration program (ETL) and reload the warehouse again. And refresh the cube. Hopefully this time, the total discrepancy for measure1 at the highest level will be 0. And we can then move on to the next measure.

Data Quality Concepts

All truths are easy to understand once they are discovered; the point is to discover them.

—Galileo Galilei, *A Single Version of the Truth*

Data must be viewed as a strategic corporate asset, and, by default, data quality must be viewed a strategic corporate responsibility. High-quality data serves as a solid foundation for enabling better business decisions for optimal business performance and ultimately business success. Data quality is one of the most important, and at the same time, most elusive aspects of an enterprise's data management efforts.

The fundamental principle that any data analyst needs to understand about data quality is that *data has to be suitable for its intended use*. It is, however, important to know the context in which the data is being used and that user satisfaction is tied directly to data quality. The foundation of data quality starts with the conformance of data, data types, and data attributes value to that domains.

Data quality is a large and complex field with many dimensions. Corporate data universe consists of numerous databases, connected by countless real-time and batch data interfaces. The data continuously moves about and changes. The databases are endlessly redesigned and upgraded, as are the programs responsible for the data integration. The typical consequence of these dynamics is that information systems get better, while data quality deteriorates. Without a comprehensive data quality monitoring program, bad data spreads like viruses.

A common definition for data quality is *fitness for the purpose of use*; the problem with this definition is that most data has multiple uses, each with its own fitness for purpose requirements, thus making data quality a multidimensional property. From that perspective, data quality definition is slightly redefined to *fitness to serve each and every purpose*. In order to achieve this data quality standard, raw data is extracted directly from its sources, profiled, analyzed, transformed, cleansed, documented, and monitored by data quality processes designed to provide and maintain the enterprise's information desires for *one single version of truth*.

Enterprise DWH provides an expellant opportunity in helping to finally archive this long desired objective. In his brilliant book, *Data Driven: Profiting from Your Most Important Business Asset*, Thomas Redman explains:

> A fiendishly attractive concept is . . . "a single version of the truth" . . . the logic is compelling . . . unfortunately, there is no single version of the truth.

> For all important data, there are . . . too many uses, too many viewpoints, and too much nuance for a single version to have any hope of success.

> This does not imply malfeasance on anyone's part; it is simply a fact of life.

> Getting everyone to work from a single version of the truth may be a noble goal, but it is better to call this the "one lie strategy" than anything resembling truth.

Looking at Ralph Kimball's concept of *Dimensional Modeling*, more specifically *Conformed Dimensions*, brings up the three most important proprieties of a data warehouse, subject-oriented, integrated, and conformed data entities that bring together all key data assets at an enterprise level. This is actually complementary with Bill Inmon's definition, stating that

> a warehouse is a subject-oriented, integrated, time-variant, and nonvolatile collection of data in support of management's decision-making process.

Interpreting both of these definitions from the business perspective leads to conclusion that enterprise DWH is providing a golden opportunity for delivering (if we do it correctly) a "single version of the truth," for an enterprise.

Discovering and understanding the single version of the truth must be the goal (holy grail) of every data analyst as data is at the heart of any modern enterprise, and that it must be used for delivering strategic business advantage.

The first step in achieving that goal is the understanding of the key data quality dimensions which contributes to data quality. Dimensions of data quality typically include accuracy, reliability, importance, consistency, precision, timeliness, fineness, value, conciseness, and usefulness.

Data Quality Dimensions

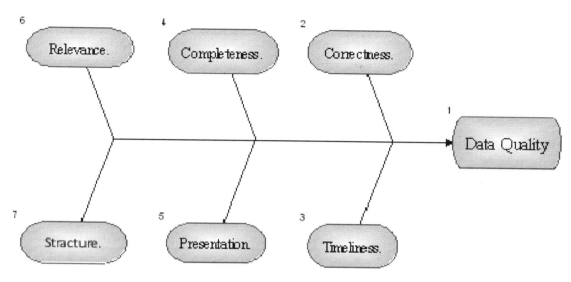

Correctness—Accuracy

Does data accurately represent the real world values they are expected to model? Incorrect spellings of product or person names, addresses, and even untimely or not current data can impact operational and analytical applications. Is the data specific—Granular specific—enough, not conflicting no missing information.

Correctness - Accuracy	
Accurate	Is the data correct?
Complete	Is there anything missing?
Precise	Is data adequately specific?
Granular	Is data sufficiently detailed?
Consistent	Is there any conflicting data?

Timeliness

Is the data up to date, enough history without historical gaps? Is it logically sequenced?

Timeliness	
Currency	Is it up to date?
Retention	Does it have enough history?
Continuous	Does it have historical gaps
Sequential	Logically sequenced?

If this database is to be used to compute sales bonuses that are due on the fifteenth of the following month, it is of poor data quality even though the data in it is always eventually accurate. The data is not timely enough for the intended use.

However, if this database is to be used for historical trend analysis and to make decisions on altering territories, it is of excellent data quality, as long as the user knows when all additions and changes are incorporated. Waiting for all the data to get in is not a problem because its intended use is to make long-term decisions.

Business Relevance

Distinct occurrences of the all data instances agree with each other or provide conflicting information. Are values consistent and useful across data sets?

Business Relevance	
Aligned	Does it match business need?
Trusted	Trust in information?
Understandable	Comprehensible?
Reliable	Available when needed?
Useful	Valuable information?

Completeness

Do data accurately represent the real world objects values they are expected to model?

Completeness	
Complete	All information available?
No Gaps	Any data value missing?
Incorrect	Some data value in unusable state?

Presentation

Presentation	
Clear	Easy to view?
Organized	Structured in logical sequence?
Nonambiguous	Not misleading
Fit-for-Tool	Easy to import/export?

Structure

Unique key, no orphan records, referential integrity, only valid values, functional constraints.

Structure	
Identity	Unique keys?
Reference	No orphan records?
Cardinality	Relationship constraints?
Dependency	Functional constraints?
Values	Only valid values?

Integrity: What data is missing important relationship linkages? The inability to link related records together may actually introduce duplication across your systems.

Motivation for DWH Data Quality

Poor data quality leads to bad business decisions. Bad business decisions leads to lost revenue. Lost revenue leads to suffering. The only thing necessary for Poor Data Quality is for good businesses to Do Nothing.

—Jedi Master Yoda

Organizations depend on data. Regardless of the industry, revenue size, or the market it serves, every organization relies on its data to produce useful information for effective business decision making. Unfortunately, with so much emphasis on such information, data quality rarely gets the attention it deserves. No one wants to consciously admit that their business decisions are based on inaccurate or incomplete data. The most important assumption to make when discussing data quality is that all conclusions of data analysis are subject to qualifications about the quality of the data. The computer acronym GIGO, standing for Garbage In, Garbage Out, applies just as much here as elsewhere, and data analysts of whatever flavor cannot perform miracles and extract gems from rubbish.

A recent survey revealed that 75 percent of organizations have no data quality processes in place. It appears that the majority of the businesses have even taken no steps to determine the severity of the problem.

Here is a question, Why is dirty data the norm rather than the exception?

The answer is that dirty data is the norm simply because we make the wrong assumption. Let's assume the data is always correct, shall we? No, we shall not, as the norm is just the opposite, unless something is done about it.

One of the root causes of poor data quality is not demanding the correct data. This section deals with causes and offers some solutions to the dirty data phenomenon.

While it's true that many organizations within an enterprise *persist* in data isolation (data silos) due to the fact that it was created while separate business entities, and at time an advantage in the sharing of its data was not a requirement, but sharing of data has become a basic requirement of any DWH project.

Data quality reliance on reactive (i.e., data cleansing processes) is focused on correcting existing data problems, but without resolving their root cause—and in some cases without even identifying it. Reactive data quality (i.e., "cleaning the lake," in Redman's analogy) is like driving by looking in the rear view mirror; it is solely focused on finding and fixing the problems with existing data after it has been extracted from its sources.

This approach to data quality has some serious disadvantage; it addresses data quality issues reactively, after data quality problem has been uncovered, and maybe upon business request or even complaint. Some data quality issues may not be easy to discover, and business users cannot decide which report is right and which one is wrong. The organization may eventually come to a conclusion that their data is bad, but it would not be able to indicate what exactly needs to be done to fix in the bad data.

The more effective proactive method is based on data profiling. A number of data profiling metrics is typically introduced to screen for missing or invalid attributes, duplicate records, duplicate attribute values that are supposed to be unique, frequency of attributes, cardinality of attributes and their allowed values, standardization and validation of certain data formats (i.e., data format) for simple and complex attribute types, violations of referential integrity constraints, etc. A limitation of the data profiling techniques is that an additional analysis is required to understand which of the metrics are most appropriate for the business and why. Definitive answers may be hard to come up with and even more difficult to translate it into a data quality improvement action plan.

Fortunately, more and more enterprises are beginning to appreciate the critical importance of viewing data as a corporate asset. This is shift toward establishing a proactive data governance program, where the data quality objective is to prevent errors at the sources where data is entered or received, *not* after data is extracted for use by DWH (i.e., "before the pollution enters the lake," in Redman's analogy). Steve Sarsfield, in his excellent book[7] explains what the data governance is:

> Data governance is about changing the hearts and minds of your company to see the value of information quality . . . data governance is a set of processes that ensures that important data assets are formally managed throughout the enterprise . . . at the root of the problems with managing your data are data quality problems . . . data governance guarantees that data can be trusted . . . putting people in charge of

---

[7]   *The Data Governance Imperative* by Steve Sarsfield

fixing and preventing issues with data . . . to have fewer negative events as a result of poor data.

This allows for the establishment of master data management system, which allows for many versions of the truth to enter, but one and only one, all-encompassing shared version of the truth to exit.

Data Quality Problems Can Slip Through at Any Stage of Development

- Data Source
- Data Integration and Data Profiling
- Data Staging and ETL
- Database Scheme (data modeling)

Quality of data can be compromised at any point, depending upon how data is received, entered, integrated, maintained, processed (Extracted, Transformed and Cleansed) and loaded. Data may be impacted by numerous processes that bring data into your data warehouse environment, most of which may affect its quality to some extent. All these phases of data warehousing are responsible for data quality in the data warehouse. Despite all the efforts, there still exists a certain percentage of dirty data. This residual dirty data should be reported, stating the reasons for the failure in data cleansing methods employed in correcting for the same. Data quality issues can arise at any stage of data warehousing. It may originate, in data sources, in data integration & profiling, in data staging, in ETL and database modeling. Following model is depicting the possible stages which are vulnerable of getting data quality problems:

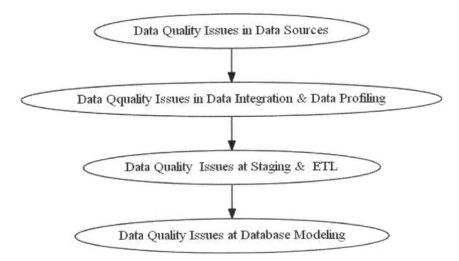

Taxonomy of the data quality issues at different stages of data warehousing

Data Quality Issues at Data Source

Data errors coming from the source file will have the most profound effect on DWH. These are the files that are coming from the systems and applications that usually outside of the enterprise DWH team control. Changes to these applications could come without any previous warning. So the continuous monitoring of the source data quality is a requirement. The most frequent failure of business intelligence project is caused by the wrong or poor quality data of the input files. Eventually, data in the data warehouse is fed from various sources as depicted in the figure below

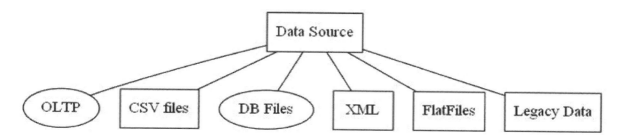

The source system consists of all the production-transaction raw data providers, from where the details are pulled out, making it appropriate for feeding at the data warehousing. All the data sources are having their own methods of extracting and storing data. Diversity of data may contribute to data quality problems, if it is not properly taken care of. Sources may be exposed to some any kind of unsecured access making them unreliable which ultimately may be contributing to poor data quality. Other data sources may have different kind of problems inherent with it, such as data from the legacy mainframe for which data may not even have metadata that describe them. The sources of erroneous data include data entry error (human or errors, incorrect data update error by a human or computer system).

Following condition may cause data quality problems with data source system:

- Inappropriate selection of candidate data sources (sources which do not comply with corporate or ANCII standards or to comply with corporate business rules).
- Misunderstanding of inter-dependencies among data sources.
- Unanticipated changes in source application.
- Multiple data sources generate semantic heterogeneity which leads to data quality issues.
- Use of different data representation formats in data sources (ASCII, EBCDIC).
- Measurement and interpretation errors.
- Using hyphens to indicate negative numbers.
- Using different data types for similar columns; for example, addressID is stored as a number in one table and a string in another.
- Specifying NULL character inappropriately in flat files.

- Metadata and data mismatch.
- Orphaned or dangling data.
- Delimiters in a file come as a valid data character in some field of a file.
- Inappropriate data entity relationships among tables.
- Updating DWH with outdated the sources data.
- Conflicting information present in data sources.
- Various representations of data in source data. (The day of the week is stored as M, or Mon, and Monday in other separate columns).
- Lack of validation routines at sources.
- Inadequate QA standards in testing data at the individual data source.
- Lack of business ownership, policy and planning of the entire enterprise data contribute to data quality problems.
- Columns having incorrect data values (i.e., The AgeInYear Column for a person contain value 3 although the birth date column is having value MMM,DD,YYYY)
- Having Inconsistent/Incorrect data formatting (The name of a person is stored in one table in the Format First-name Surname and in another table in the format "Surname, First-name").
- Missing or additional columns in the source file.
- Missing or misspelled values in data sources.
- Multiple sources for the same data presented in a different format.
- Inconsistent use of special characters (i.e., a date may different special characters to separate the year, month, and day.

Data Quality Issues When Staging Data during ETL

Before we start looking into data quality problems at this stage, we need to know how the ETL system handles data rejection, substitution, correction, and notification without modifying data. To ensure success in testing data quality, include as many data scenarios as possible. Typically, data quality rules are defined during design, for example:

- Reject the record nonnumeric data if it is a numeric field.
- All source data that is expected to get loaded into target actually gets loaded.
- All fields are loaded with full contents—i.e., no data field is truncated while transforming.
- No duplicates are loaded.
- Substitute null if a decimal field has nonnumeric data.
- Aggregations take place in the target properly.
- Verify and correct the state field if necessary based on the ZIP code.
- Data integrity constraints are properly taken care.
- Verify results based on the business rules.

For example, the business rule is "if there is no match found in the lookup table, then load it anyway, but report the error to a user in the error log/report."

Depending on the data quality rules of the application being tested, scenarios to test might include null key values, duplicate records in source data and invalid data types in fields (e.g., alphabetic characters in a decimal field) this may be the issues with source data. For the purpose of continuous quality improvement efforts, it may be beneficial to create a quality database and populate the *before* data in this database for users and the QA team to view to be able to take appropriate actions.

A data cleaning process is executed in the data staging area to improve the accuracy of the data warehouse. The data staging area is the most appropriate place where all *grooming* is done on data after it is culled from the source systems.

The staging and ETL phase is considered to be most crucial stage of data warehousing where the full responsibility of data quality efforts resides. It is a prime location for validating data quality from source or auditing and tracking down data issues. There are many factors for data quality problems at this phase, some of them are listed below.

- Nonrelational type staging area.
- Conflicting business rules of various data sources.
- The inability to schedule extracts on time, or within the allotted time interval.
- Inability of capturing only changes in source files.
- Disabling data integrity constraints in staging tables cause wrong data and relationships to be extracted.
- Lack of effective and centralized metadata repository.
- Lack of replication of data cleansing of rules established for cleaning data, into the metadata.
- Misinterpretation of the slowly changing dimensions strategy in ETL.
- Inappropriate transformation of null values in ETL process.
- Absence of automated unit testing facility in ETL tools.
- Lack of effective error reporting, validation, and metadata updates in ETL.
- Inappropriate ETL process on data update strategy.
- Failure to incorporate business rules into the transformation logic.
- Loss of data during the ETL process (rejected records, refused data records in the ETL process).
- The inability to restart the ETL process from checkpoints without losing data.
- Lack of automatic rules generating ETL tools based on mappings that detect and fix data defects.
- Inability of integrating profiling, cleansing, and ETL tools for reconciling data and metadata.
- Inability to cope with ageing data.
- Varying timeliness of data sources.

- Miss-aligned key strategies for the same type of entity (i.e., One table stores customer information using the Social Security Number as the key, another uses the CustomerID as the key, and another uses a surrogate key).

Review the detailed test scenarios with business users and technical designers to ensure that everyone is in agreement. Data quality rules applied to the data will usually be invisible to the users once the application is in production; users will only see what is loaded to the database.

For this reason, it is important to ensure that what is done with invalid data is reported to users. These data quality reports present valuable data that sometimes reveals systematic faults.

Metadata

Until recently, a neglected but very important component of any database—but even more so of the DWH—is metadata (data about data). Metadata and data heredity provide the business definitions of the data, the technical specifications of the data, and a process that shows how the data has been processed, what business rules are applied to the data, and what validation criteria were applied to the data. In the new world of DWH, metadata has taken a higher level of importance, and it has become imperative for it to be part of any DWH. Without it, data analysts, programmers, and others would be poking into DWH forever without any guarantee of success. Metadata is like an index into DWH as it stores the following information:

- Structure of the data from the programmer's point of view.
- Structure of the data from the data analyst's perspective.
- Sources of data feeding DWH.
- Business rules for data transformation.
- Data model.
- Relationships of data.
- History of extractions.

Zachman's Architecture Framework

The notion of metadata goes back to the Zachman's Architecture Framework.

The Zachman's Framework consists of a matrix in which the rows represent the perspectives of different people (stakeholders) have on an IT project, and the columns represent what they are seeing from that perspective. From the perspective of metadata here, we are only interested in the first *data* column of the Zachman's Architecture. For the purpose of explaining the famous Zachman's Architecture

Framework and contact of our data column in it, a short description is provided here, but for more information, please see URL referenced below.

The Zachman framework solves the complexity of the enterprise architecture by decomposing the problem into a two dimensional matrix, with each dimension consisting of varicose subcategories. The first dimension (rows of the matrix) defines the perspective of one of category of players in enterprise system development process, such as:

- *Scope* (Planer's view): Enterprise's direction and purpose-context.
- Business model (Business owner's view): Business concepts, structure, processes, organization, and so forth.
- *System model* (Architecture view-logical): Specific organization, computer systems, forms procedure required to carry out a business.
- *Technological model* (Builder's view): Describes how technology may be used to resolve information-processing needs of the organization.
- *Detail representation* (Sub-contractor's view): Describes details such as database, programming language, network, and so forth.
- *Functioning system* (Inventory view): Describes actual computer system installed along with their database.

The second dimension, which is represented in the column of the matrix, consists of key decision driving questions: What, how, where, who, when and why.

- *What* (column 1) describes what data flow throughout the enterprise
- *How* describes functions and processes performed by different systems.
- *Where* describes networks that provides interconnections for information delivery.
- *Who* specifies people and affected by the architecture.
- *Why* describes business drivers for the architecture.
- *When* defines timing constraints in relation to processing requirements.

The strength of this architecture is the notion of architecture and design patterns. Architectural patterns refer to an approach to a solution that has been successfully implemented number of times.

Zachman's Architecture Framework matrix is reprinted on the next page with permission by John A. Zachman, www.Zachman.com.

ENTERPRISE ARCHITECTURE - A FRAMEWORK ™

	DATA What	FUNCTION How	NETWORK Where	PEOPLE Who	TIME When	MOTIVATION Why	
SCOPE (CONTEXTUAL) *Planner*	List of Things Important to the Business Entity = Class of Business Thing	List of Processes the Business Performs Function = Class of Business Process	List of Locations in which the Business Operates Node = Major Business Location	List of Organizations Important to the Business People = Major Organizations	List of Events Significant to the Business Time = Major Business Event	List of Business Goals/Strat Ends/Means = Major Bus. Goal/Critical Success Factor	**SCOPE (CONTEXTUAL)** *Planner*
ENTERPRISE MODEL (CONCEPTUAL) *Owner*	e.g. Semantic Model Ent = Business Entity Reln = Business Relationship	e.g. Business Process Model Proc = Business Process I/O = Business Resources	e.g. Business Logistics System Node = Business Location Link = Business Linkage	e.g. Work Flow Model People = Organization Unit Work = Work Product	e.g. Master Schedule Time = Business Event Cycle = Business Cycle	e.g. Business Plan End = Business Objective Means = Business Strategy	**ENTERPRISE MODEL (CONCEPTUAL)** *Owner*
SYSTEM MODEL (LOGICAL) *Designer*	e.g. Logical Data Model Ent = Data Entity Reln = Data Relationship	e.g. Application Architecture Proc = Application Function I/O = User Views	e.g. Distributed System Architecture Node = I/S Function (Processor, Storage, etc) Link = Line Characteristics	e.g. Human Interface Architecture People = Role Work = Deliverable	e.g. Processing Structure Time = System Event Cycle = Processing Cycle	e.g. Business Rule Model End = Structural Assertion Means = Action Assertion	**SYSTEM MODEL (LOGICAL)** *Designer*
TECHNOLOGY MODEL (PHYSICAL) *Builder*	e.g. Physical Data Model Ent = Segment/Table/etc. Reln = Pointer/Key/etc.	e.g. System Design Proc = Computer Function I/O = Data Elements/Sets	e.g. Technology Architecture Node = Hardware/System Software Link = Line Specifications	e.g. Presentation Architecture People = User Work = Screen Format	e.g. Control Structure Time = Execute Cycle = Component Cycle	e.g. Rule Design End = Condition Means = Action	**TECHNOLOGY MODEL (PHYSICAL)** *Builder*
DETAILED REPRESENTATIONS (OUT-OF-CONTEXT) *Sub-Contractor*	e.g. Data Definition Ent = Field Reln = Address	e.g. Program Proc = Language Stmt I/O = Control Block	e.g. Network Architecture Node = Addresses Link = Protocols	e.g. Security Architecture People = Identity Work = Job	e.g. Timing Definition Time = Interrupt Cycle = Machine Cycle	e.g. Rule Specification End = Sub-condition Means = Step	**DETAILED REPRESENTATIONS (OUT-OF-CONTEXT)** *Sub-Contractor*
FUNCTIONING ENTERPRISE	e.g. DATA	e.g. FUNCTION	e.g. NETWORK	e.g. ORGANIZATION	e.g. SCHEDULE	e.g. STRATEGY	**FUNCTIONING ENTERPRISE**

Relating Data Columns with Metadata

Let's take a short diversion that will help understand why and what kind of information about data should be kept in metadata repository. Let us go back to our data column of the Zachman's matrix. The data column of the architecture framework is concerned with *what* is the importance, from each of the six points of view to an enterprise:

- *Row 1—The planer* concerns are the aggregates of the similar groups of data.
- *Row2—The business owner* conceptualizes behind the business and the facts that link them.
- *Row 3—The architect* is concerned with conversion of business concepts to structure suitable for information processing. The architect is concerned with database schema.
- *Row 4—The builder* (designer) uses the architect's structure to define the how data management technology is used to support it. The builder concerns are rational tables and columns.
- *Row 5—Subcontractor* (implementer) loads data into the storage medium.
- *Row 6—Functional system* is an inventory of physical databases.

The above list represents the hierarchy of information from the each of the six points of view of each, and it should be stored and kept up to date on the metadata repository.

Metadata is just repository for computerized database containing metadata. Number of vendors offer metadata repository, consisting of an empty database and a tool for manipulating metadata Mostly these metadata implementation start our correctly and eventually get out of sync with data in the DWH, therefore become useless.

One of the fundamental concepts of data warehousing is integrating data. DWH provides the opportunity of achieving consistent error-free information. The error-free integration may only be achieved by reconciliation process in metadata repository depicted in the diagram blow:

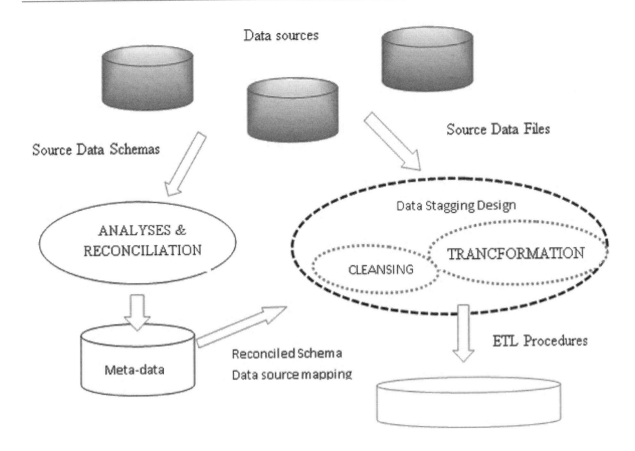

Metadata is created by analysis and reconciliation. Metadata, in turn, reconciles mapping between reconciled and the source schema attributes.

An Introduction to Data Profiling

Data profiling may be the most important data quality effort that can be employed by the IT data warehouse development team. According to data warehousing professionals, the top two challenges in implementing ETL tools are ensuring adequate data quality and understanding source data. Most data warehousing professionals learn the hard way that fixing unanticipated data defects and anomalies in the source data (and in some cases, data warehouse data) is the primary reason for project delays, overruns, and failures.

Despite the growing publicity about the impact of poor quality data on data integration initiatives, most organizations take only minimal steps to understand—or *profile*—the data they want to integrate. Most organizations report data quality problems *after the fact*—when users complain about errors in the new system, according to a Data Warehouse Institute survey.

Problem databases cost corporations millions of dollars annually. They consume enormous amounts of IT budgets when attempting to use them in new projects. They cause loss of money and time through inaccurate data and incorrect use of accurate data.

Problem databases exist in almost all corporations. They are not small, insignificant databases; they are usually the mainstream operational databases that the corporation runs on.

Working with problem databases is very frustrating. They cause excessive delays in completing projects. They cause new quality problems in derivative databases. Information system professionals get frustrated when working on projects that seemingly never end. Management gets frustrated at not being able to predict the time needed to complete projects, the eventual cost or the probability that the outcome will be what they expect.

What Is a Trouble-prone Database or Data Warehouse?

A database presents major issues when it lacks the ability to be changed or used in new ways, without a great deal of difficulty. They can be poorly documented, inaccurately documented, have unknown or questionable data quality, be difficult to extract data from because of structural problems, or be difficult to interface with.

Some of the characteristics that make a database a trouble-prone database are as follows:

- No metadata descriptions available for designers, developers, or troubleshooters
- Metadata documentation that does not match the data
- Data elements used for different purposes than documented
- Program descriptions of data that are not correct
- Data that is inconsistently encoded within a data element
- Incorrect data values based on existing documentation or specs
- Missing data values when they should exist
- Ad hoc and undocumented methods for expressing null or not applicable (n/a)
- Overloaded data elements: more than one business fact included in the same element (ex. account name and number in same field)
- Denormalized data structures that are not documented as being denormalized
- Violations of referential constraints (orphaned values, orphaned foreign keys)
- Data rule violations indicating incorrect data values
- Nonunique primary keys
- Functional dependency violations across tables
- Duplicate data in fields or entire records

The common theme through all of this is inaccurate data and metadata. Inaccurate metadata leads to incorrect use of data. Most of the problems are not known. They are waiting to be discovered, and will be, at the most inopportune times.

If one were to score a database on these factors giving each item on the list a value of 0-10 where 10 is bad, and add up the numbers, you could label a database troublesome at some threshold, for example, 50. Not many databases would fall below this line. A database can be a problem database with only one of the negative factors being present. For example, inaccurate data can render a database almost useless if it is prevalent enough.

Data profiling is an analysis of the candidate data sources for a data warehouse to clarify the structure, content, relationships, and derivation rules of the data. Data profiling should be considered for source data and data that has been loaded to the data warehouse staging, dimension, and fact tables, among others, to verify that data quality is built into the new tables.

Data profiling helps developers and other project stakeholders not only to understand anomalies and to assess data quality, but also to discover, register, and assess enterprise metadata. Thus, the purpose of data profiling is both to validate metadata when it is available and to discover metadata when it is not. The result of the analysis is used both strategically—to determine suitability of the candidate source systems and give the basis for an early go/no-go decision, but tactically, to identify problems for later solution design, and to level sponsors' expectations.

Data profiling results (information and statistics) may be used for the following purposes.

1. Determine whether existing data can be used for required purposes.
2. Provide metrics on data quality, including whether the data conforms to particular standards or patterns, and where it does not aid in identifying the problems with the data.
3. Assess the risk involved in integrating data for new applications, including the challenges of table joins.
4. Assess whether metadata accurately describes the actual values in the source data.
5. Understand data challenges early in data intensive projects so that late project surprises are avoided. Finding data problems late in the project can lead to delays and cost overruns.
6. Provide an enterprise view of all project data for uses such as master data management where key data is needed and data governance for improving data quality.

Highlights of Data Profiling

Data profiling utilizes different kinds of descriptive statistics such as minimum, maximum, mean, mode, percentile, standard deviation, frequency, and variation, as well as other aggregates such as count and sum. Additional metadata information obtained during data profiling may be data types, field lengths, discrete values, uniqueness, occurrence of null values, typical string patterns, abstract type recognition, and more. The metadata can then be used to discover problems such as illegal values, misspelling, missing values, varying value representation, and duplicates.

Unique analyses are performed for different structural levels. For example, single columns can be profiled individually to get an understanding of frequency distribution of different values, type, and use of each column. Embedded value dependencies can be exposed in cross-columns analysis. And overlapping value sets possibly representing foreign key relationships between entities can be explored in an inter-table analysis. Normally tools are used for data profiling to ease the process. The computation complexity increases when going from single column, to single table, to cross-table structural profiling. Therefore, performance is an evaluation criterion for profiling tools.

Data Profiling Methods

In general, there are three methods employed for data profiling.

1. *Column profiling* (the most commonly used) provides statistical measurements associated with the frequency distribution of data values (and patterns) within a single column (or data attribute). Resulting metadata can identify the number of records, the number of null (i.e., blank) values in each field, data types (e.g., integers, characters), field length, unique values, patterns in the data, relationships to other fields, etc. By first evaluating metadata, an analyst gains a general sense of the cleanliness and structure of the data and pinpoints potential problem areas.

2. *Cross-column profiling* analyzes dependencies among sets of data attributes within the same table. Analysts may want to understand the relationships of records within or between tables. For example, an analyst might validate a primary/foreign key relationship and discover that there are many customers who have purchased products but haven't been billed for them because their customer numbers don't exist in the billing system.

3. *Data rule validation* verifies the conformance of data instances and data sets with predefined rules. Analysts will want to analyze actual data values. This will help determine whether the data adheres to business rules or not. If the data is deemed invalid, the business analyst knows he will need to devise rules to transform the data so that it meets corporate standards.

In the table below you will see several types of profiling that can be useful.

Types of Profiling	Example	Types of Data Profiled
Range Checking	Commission Amount: identify min, max	All fields with numerical and alphabetical values
Domain Values	Transaction Detail Type: WR, SO, SR, BR, BD, AB	All fields defined with acceptable values in mapping document
Cross Field Verification	RepCode is correctly related to the reps BranchCode	Fields in child tables contain data related to parent table
Field Format Verification	BusinessDate, ProductID	Fields have defined format (e.g., Date)
Reference Field Verification	Any field populated by means of ETL lookups	Fields populated via reference lookup tables

Types of Profiling	Example	Types of Data Profiled
Format Consolidation	BusinessDate, all other date fields	Fields transformed to meet common format requirement
Defaults Applied		Where source field is null, rules indicated population with default value in target
Referential Integrity	Foreign key in Transaction Detail table points to valid primary key in Transaction table.	Valid foreign key to primary key association
Outliers	AdjustmentAmount, Adjustment Detail Rate	Codes, dollar values, etc. do not exceed expectations
Ranges	Date ranges reasonable, correct	Dates, amounts
Duplicates	Duplicate records, duplicate transactions, duplicate reps	Records, dates
Uniqueness, Distinct	Transactions, broker dealer	Where transactions, reps, etc. should be unique and not repeated
Business Rule Compliance	Using mapping document business rule descriptions	Transformed data with business rules applied
Missing Data	Not null fields are null	Data fields which should be not null

This is just an introduction to profiling. A good understanding of the benefits and methods should be considered by anyone doing data warehouse and ETL testing. Two of our favorite books are listed below. Not many others are available.

Jack E. Olson, *Data Quality, the Accuracy Dimension*, Morgan Kaufmann, 2003.
Arkady Maydanchik, *Data Quality Assessment*, Technics Publications, 2007.

Assessing Data Profiling Tools

Data profiling with vendor tools has advantages over manual methods. Most users profile data by unpredictably running ad hoc queries to view table column contents and recording, in word processing files, what the queries revealed. This amasses documentation that's seldom updated and difficult to apply directly to a data integration or data quality solution. This practice is neither comprehensive nor methodical, so users may miss important data and relationships across data structures. Therefore, manual methods of this type are inferior, because of the time-consuming and error-prone process of moving profile information from a query to the documentation to the data management solution.

When possible, users should profile with a vendor tool (whether it's an autonomous tool or a collection of functions within a larger tool) to gain greater accuracy, completeness, repeatability, and productivity.

Another way to gain insight into data is to write queries, typically SQL queries, against database tables. There are a couple of problems with this approach. First, data from multiple sources—often a large number of sources, millions of rows of data, and diverse formats—must be consolidated. The data can be in flat files or DB2 files, in ASCII or UNICODE, on UNIX systems or mainframes. The mere act of consolidating these sources can present a significant IT challenge. The second problem is the unsatisfactory results of traditional querying under these circumstances. Without a clear understanding of how the data is interrelated within a data set and how sets are related to each other, it is difficult to know what queries to run. Not knowing what to look for makes it hard to zero in on the inconsistencies and anomalies that need to be addressed. Often this method of querying creates more questions than it answers.

Profiling Tools Typically Provide:

- Extraction/specification/adjustment of basic metadata.
- Statistical analysis of column values and relationships.
- Repository holding metadata and analysis results.
- Drill down to the actual records.
- Ability to document the business rules.
- Ability to attach text notes for questions and/or answers.
- Manipulation to *normalized* or *target*.
- DDL generation.

The figure below shows where to use data profiling in the development process.

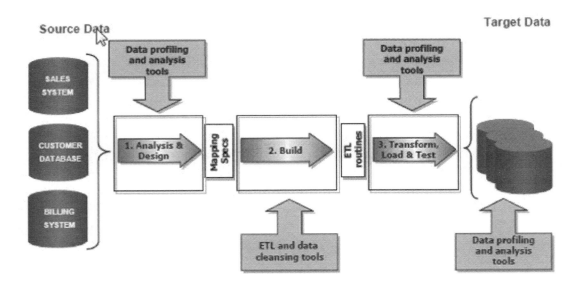

Now that you better understand how to analyze, clean, and monitor data, you may wonder which data profiling tools or manual methods to purchase. The good news is that today there are more data profiling tools on the market than several years ago. However, not all data profiling tools work the same way, so you will need to carefully evaluate their capabilities.

Ideally, you should prototype a tool by running it against a subset of your data that you know well. This will help you understand whether the tool captures key characteristics in the data and how easy it is to use.

To help you in your quest, here are a few guidelines to help you find the data profiling tool that is right for your organization.

- *Ease of use.* Can a business analyst use the tool without having to rely on someone from IT to run the software and generate the reports? The use of wizards, side-by-side panels that display summary and detail data, and visual workflows may indicate that the tool is appropriate for business users.
- *Collaboration.* Good profiling tools educate the wider community about the nature of data that comprises a new or existing application. That means the tools should be able to generate easy-to-read reports in various formats (e.g., HTML, Word, Excel) that quickly communicate data quality issues to relevant constituents.
- *Direct Connectivity to Sources.* To ensure adequate scalability, data profiling tools should connect directly to source systems rather than create copies of the data, which may take hours or more and chew up valuable processing power. Tools that connect directly to data sources also ensure that analysts are accessing the most up-to-date data values.
- *Generate Rules for Cleansing.* It is not enough to identify data issues. Analysts must be able to create rules to fix the problems. Ideally, analysts should be able to create rules within a data profiling tool, which can then pass the rules to a data cleansing tool for processing. In other words, data profiling tools should be tightly integrated with data cleansing tools to make the most efficient use of analysts' time.
- *Integration with Third-Party Applications.* To automate routine data integration or migration processes, such as data warehousing loads, data profiling tools (and data cleansing tools) should be integrated with third-party applications, such as ETL and data integration tools. The profiling and cleansing jobs should be initiated and managed by these third-party applications.
- *Offers Broad Functionality.* Data profiling tools should offer a comprehensive set of functionality for analyzing data structures, generating statistics about column values, and mapping dependencies within or among tables. It makes efficient use of analysts' time.

When creating or updating a data profile process, start with basic column-level analysis such as:

- *Distinct count and percent:* Analyzing the number of distinct values within each column will help identify possible unique keys within the source data (which I'll refer to as natural keys). Identification of natural keys is a fundamental requirement for database and ETL architecture, especially when processing inserts and updates. In some cases, this information is obvious based on the source column name or through discussion with source data owners. However, when you do not have this luxury, distinct percent analysis is a simple yet critical tool to identify natural keys.
- *Zero, blank, and NULL percent:* Analyzing each column for missing or unknown data helps you identify potential data issues. This information will help database and ETL architects set up appropriate default values or allow NULLs on the target database columns where an unknown or untouched (i.e., NULL) data element is an acceptable business case. This analysis may also spawn exception or maintenance reports for data stewards to address as part of day-to-day system maintenance.
- *Minimum, maximum, and average string length:* Analyzing string lengths of the source data is a valuable step in selecting the most appropriate data types and sizes in the target database. This is especially true in large and highly accessed tables where performance is a top consideration. Reducing the column widths to be just large enough to meet current and future requirements will improve query performance by minimizing table scan time. If the respective field is part of an index, keeping the data types in check will also minimize index size, overhead, and scan times.
- *Numerical and date range analysis:* Gathering information on minimum and maximum numerical and date values is helpful for database architects to identify appropriate data types to balance storage and performance requirements. If your profile shows a numerical field does not require decimal precision, consider using an integer data type because of its relatively small size. Another issue which can easily be identified is converting Oracle dates to SQL Server. Until SQL Server 2008, the earliest possible datetime date was 1/1/1753 which often caused issues in conversions with Oracle systems.

With the basic data profile behind you, conduct more advanced analysis such as:

- *Key integrity:* After your natural keys have been identified, check the overall integrity by applying the zero, blank, and NULL percent analysis to the data set. In addition, checking the related data sets for any orphan keys is extremely important to reduce downstream issues. For example, all customer keys from related transactions (e.g., orders) should exist in the customer base data set; otherwise you risk understating aggregations grouped by customer-level attributes.

- *Cardinality:* Identification of the cardinality (e.g., one-to-one, one-to-many, many-to-many, etc.) between the related data sets is important for database modeling and business intelligence (BI) tool set-up. BI tools especially need this information to issue the proper inner- or outer-join clause to the database. Cardinality considerations are especially apparent for fact and dimension relationships.
- *Pattern, frequency distributions, and domain analysis:* Examination of patterns is useful to check if data fields are formatted correctly. As example, you might validate e-mail address syntax to ensure it conforms to user@domain. This type of analysis can be applied to most columns but is especially practical for fields that are used for outbound communication channels (e.g., phone numbers and address elements). Frequency distributions are typically simple validations such as *customers by state* or *total of sales by product*, and help to authenticate the source data before designing the database. Domain analysis is validation of the distribution of values for a given data element. Basic examples of this include validating customer attributes such as gender or birth date, or address attributes such as valid states or provinces within a specified region. Although these steps may not play as critical a role in designing the system, they are very useful for uncovering new and old business rules.

Choosing the best profiling techniques depends on the project objectives. If you're building a new database from scratch, take the time to execute and review outcomes of each of the above bullet points. If you're simply integrating a new data set into an existing database, select the most applicable tasks that apply to your source data.

Improving Data Quality with Data Profiling

The importance of data profiling is difficult to overstate, as it has become the starting point of any data quality initiative. Data profiling is the most important defect prevention tool, a gateway for *preventing* propagation of defects before their entering into DWH. Data profiling has emerged as an essential technical discipline that precedes many other data warehousing operation. It encompasses data quality assessment and verification, metadata management, ETL processing, data migration and consolidation, and many others.

Understanding data is the essential foundation to effectively improving data quality. Traditional data analyses are simply unable to deal with the scope of data management problems. The answer to this challenge is data profiling, which offers the accurate and often automated method to understand large amounts of data. Profiling enables team of QA and business analysts to quickly understand and analyze the large and complex amount of data. This is especially true in dealing with the source or client's data over which they have little or no authority.

Data Quality Solution Offered by Data Profiling

THE BASICS OF DATA PROFILING

Data profiling consists of multiple analyses to investigate the structure and content of data and make inferences about data.

Column Examination

- Identify all values in column along with frequency of occurrence.
- Identify min and max values.
- Determine true data type.
- Determine degree of uniqueness.
- Determine encoding patterns used, frequency of each pattern.
- Compute values: AVG, SUM, MEDIAN, STD DEVIATION.

Row Examination

- Find all primary key candidates (single or multi-column).
- Find intra-row column dependencies (find denormalization instances).
- Find multi-column value relationships.
  - Value ordering rules
  - NULL value dependencies

Multi-table Examination

- Find matching columns across tables.
- Match by column name, data type.
  - Match by values
  - Find primary/foreign key pairs (single and multi-column)
- Determine 1-1, 1-M, 1-0, M-1, M-M, 0-1 rules.
- Find primary values not found in secondary tables.

Invalid Values

- Missing values when should not be missing.
- Values out of range or not in domain of expected values.
- Value in one column not possible when combined with values in one or more other columns.
- Example: obviously wrong values
  - Name = Donald Duck
  - Address = 1600 Pennsylvania Avenue

Examples of problems easily uncovered through data profiling analysis:

- Data elements used for purposes other than they should be.
- Empty columns; columns containing no data at all.
- Invalid values in columns.
- Inconsistent methods of representing the same value.
- Missing values.
- Violation of structural dependencies.
- Violation of expected column relationships missing date values.
- Violation of business rules.
- Unrealistic percentages of specific values appearing in a column.

Data profiling is an organized methodology for analyzing the data in stages that provides for a thorough result. The stages that an analyst typically exercises are as follows:

- Analyze individual values to determine if they are valid values for a column.
- Analyze all the values in a column together to find problems with unique rules, consecutive rules and unexpected frequencies of specific values.
- Analyze structure rules governing functional dependencies, primary keys, foreign keys, synonyms and duplicate columns.
- Validate data rules that must hold true with a row of data.
- Validate data rules that must hold true over all rows for a single business object.
- Validate data rules that must hold true over collections of a business object.
- Validate data rules that must hold true between collections of different types of business objects.
- Verify ETL Rules/Transformation logic.

Data rules are a subset of business rules that define relationships between sets of columns or rows that must always be true within the data. A violation may mean that data inaccuracies exist in the data or that the business rules they are based on are not being followed in the real world. In one case, the data was entered inaccurately. In the other case, the data was entered correctly, but the transaction was handled with data outside of the corporation's business policies. Both of these situations are important to expose.

Examples of data rules are as follows:

- Employees must be at least eighteen years old.
- Part-time employees are paid hourly.
- Checkout periods for tools cannot overlap for the same tool.
- Customers with more than $50,000 in sales last quarter get a five-percent discount.
- Suppliers cannot supply radioactive part numbers unless certified.

Potential Sources of Data Errors

Individual's name can be represented in more than one way in different data sources, so the parsing algorithm is a fair game for the verification. Here are potential problems (test cases) with individual name for the parser that needs to be verified with business analyst:

Name	Name Prefix	First Name	Middle initial	Last name
Mr. John J Smith	Mr.	John	J	Smith
Samantha Smith		Samantha		Smith
Smith, Jeff D & Corie		Jeff Corie	D	Smith

Address is the other field that often may be a source of confusion. Address parsing algorithms can easily resolve a simple address, 500 Johns Avenue. But what about the permutations presented by various data sources:

1. 500 S Johns Avenue Floor 55
2. 500-55 South Jones
3. Jones #500 55th

Street number	Predirection	Street Name	Street Type	Address Extension	Address Ext. Number
500	S	Johns	Ave	Floor	55
500	South	Johns		Unit	55
500		Johns		Floor	55

How to improve Data Quality with Data Profiling?

A frequently used technique—*column profiling*—is presented below to illustrate the power of the data profiling tool to QA teams. Column profiling provides the statistical measurements related to frequency distribution of data values within a single column. Column profiling often provides the first step in understanding data. This is the area where manual data analysis may focus attention. Unlike manual testing, data profiling methods provides additional advantages that can not be achieved manual testing:

- Ability to process entire data source rather than a subset of rows.
- Data profiling tools generate value frequency distribution lists allowing for quick review of range and spread of data values thus identifying unexpected anomalies in data.

- Generation of patterns in data such as in, zip code, or SIN, enabling quick identification of nonstandard values in data.

Column profiling allows for looking at each data attribute to evaluate basic statistics about it such as, percentage population, uniqueness, value renege, data type, field length, etc. This technique provides interesting analyses such as included in the table below.

Method	Description
Range analyses	Verifies if the values are within the range of predefined values.
Uniqueness	Verifies that each value assigned to the attribute is unique.
Value Distribution	Normal, binary, random or uniform value distribution.
Missing [Null] Values	Count number of null values.
Format Evaluation	Identifies the unrecognized format.
Minimum Value	Identifies minimum values.
Maximum value	Identifies maximum values.
Mean	Calculated the average for value for the numeric data.
Median	Calculates middle value.
Standard Deviation	Calculate standard deviation for numeric data.
Completeness	Percentage of populated values.
Cardinality	Enumerate the number of distinct values.
Sparseness	Evaluates the percentage of value that are not populated.
Actual count of values	Number of populated values.

Let us use an example of column profiling to illustrate how testers would use the profiling tool. Data profiling tools may automate some of the mundane work but there is no substitute for the analyst who understands data.

Primary key analysis

Primary key analysis information allows validating the primary keys that are already defined in the data and identify columns that are candidates for primary keys.

Primary keys columns have a high percentage (approaching 100 percent) of uniqueness, as well as at least one foreign key candidate. There is only one primary key per table. For example, a customer table with a customer ID that was auto-generated and other tables in the data source link to that customer ID, the customer ID is a good candidate for a primary key. It is unique and was auto-generated for the purpose of being a primary unique identifier.

The table below represents the partial output of the profiling tool:

Field name	Null	Missing	Actual	Completeness	Uniqueness
Customer_ID	0	0	5,850,290	100%	100%

Typically profiling tools will allow drilling-down on any field by clicking on it would display a field summary for that field. Good profiling tools are connected to a metadata repository. Metadata contains information that indicates the data type and field length. It also indicates if a field can be missing or null or if it should be unique. In the case of the Customer_ID, it displays data type (integer in this case) and field length. The data profiling tool provides a snapshot of the basic statistics for an entire data field by presenting statistical information such as minimum and maximum values, mean, median, mode, and standard deviation, to highlight aberrations from normal values statistics for *Customer ID.*

The profiling tool can be used to drill-downs on the field to get more details about *Customer ID* such as frequency distribution of the numeric patterns to include number of 10 digits, 9 digits, 2 digits, etc. Based on this information from the profiling tool, it can be safely concluded that the *Customer ID* is a unique integer surrogate key that can be used as the primary key for this data source.

Gender Code field has 6 distinct values, as per below:

Field name	Null	Missing	Actual	Completeness	Cardinality
Gender_Code	398,390	59,800	4,850,290	83%	8

The profiling tool after drilling down will provide the distribution of values for *Gender Code* as per the table below:

Gender Code (pattern)	Count	Percentage
1	90,389	13%
M	150,233	22%

Male	110,358	16%
2	80,000	11%
F	100,900	15%
Female	66,000	9%
Null	7,000	10%
Unknown	2,000	2%

It may be obvious to conclude that *F* is an abbreviation for *Female* and *M* is an abbreviation for *Male*, but it is dangerous to assume anything with the mimetic value (1 and 2). The *Unknown* may be used when the name is nether *Male* nor *Female*.

Something else that may be concluded from this field is that it needs further consolidation into three groups (Male, Female, and Organization).

Pattern matching: Typically pattern matching is used to verify the data values in a field are in expected format and that this information is consistent across all data sources. To demonstrate this let us take an example of North American telephone numbers. Valid format of telephone number consists of three sets of numbers (three numbers for the area code, three numbers for the exchange, and four numbers for station). These sets may or may not be separated by special characters or blank. In this example, *9* represents any digit, *X* represents any alpha character. Valid pattern could be the following:

- 999-999-9999
- (999)999-9999
- 999-999-XXXX
- 999 999 9999

The profiling tool after drilling down will provide the distribution of values for the telephone number as per the table below:

Telephone number (pattern)	Count	Percentage
(999)999-9999	90,389	13%
999-999-9999	150,233	22%
(999)-999-9999	110,358	16%
999 999 9999	80,000	11%

999-999-XXXX	100,900	15%
9-999-999-9999	66,000	9%
99 99 999 9999	7,000	10%
X	2,000	2%

The information is provided from the metadata such as:

- Data Type: VARCHAR, and
- Data Length: 25

Profiling tool provides additional statistics such as:

1. Minimum Field Length: 1
2. Maximum Field Length: 13
3. Minimum Field Value: 0-000-000-0000
4. Maximum filed value: 9-999-999-9999

Completeness value provided by the profiling tool may be misleading as it counts in invalid values as the actual values.

Frequency counts and outlier detection In essence, this technique highlights the data that nee further investigation. This is illustrated below with field name *State Abbreviation*.

Field name	Null	Missing	Actual	Completeness	Cardinality
State Abbreviation	156,899	0	4,850,290	79%	69

It appears that the *cardinality* for the *State Abbreviation* is the suspect because, if we assume that the content of cardinality matches its metadata, then only 62 distinct values for US states should be expected. Drilling-downs on this field reveal, from metadata, that data type is CHAR, and the field length is 2. Therefore, *State Abbreviation*, when populated, should contain two characters, but profiled field lengths for minimum is 2 and for maximum is 10. Let's take a look at profiler's display below:

State Abbreviation	Count	Percentages
CA	390,899	31.9%
OH	300,789	24.6%

TX	290,876	23.8%
OH	20,235	1.7%
NY	130	10.6%
CO	50	4.1%
CT	20	1.6%
Invalid value	11	0.9%
California	5	0.4%
Ca	2	0.2%
Aalabama	1	0.2%

Problems such as incorrect states spelling or typos (Aalabama), multiple state representations (California is represented as "CA," "Ca," and "California"), etc., become obvious after being deployed in a drilled-down panel. Outlier detection allows finding a few values that are remarkably different from the others (i.e., misspelled Alabama)

Multi-column value relationships (dependency)

The focus of this test is to verify column relationships in a single table. For example, if values a and b (i.e., a+b) from two fields (u, w) should always populate three other columns (x, y, z), then this correlation should always be true for all the rows, and it will always be true for all permutations of (x, x, y), or (x, y, y) as well. This implies that value in columns X, Y and Z are predetermined by values from columns U and W.

For example, the postal code in Canada should correspond to two other columns, province, and the city name; if it populates something else, it should be investigated further. For US State Abbreviation field, the profiler may show two or more values in the Zip Code field for the same State Abbreviation field.

Join Testing

This is an essential step in testing integrated data from separate sources. In data warehouse scenarios, this can be envisaged to create a single view of customer data. Data in the DWH is drawn from previously unconnected sources managing accounts, customer relationships, ordering, delivery, etc. Any assumption that is used in merging data must be tested. The principal consideration in join testing is the relationship discovery to verify that data sources at properly aligned, for example:

- Is more them one relationship across tables?
- Is primary/foreign key relationship enforced?
- Product ID exists in the Invoice panel, but there is no such product in the product database.
- Customer ID exists on the sales order but no corresponding customer or has multiple customer records with the same Customer ID in the customer database.

Cross-domain analysis

Cross-domain analysis examines content and relationships across tables. This analysis identifies overlaps in values between columns and any redundancy of data within or between tables. For example, country or region codes might exist in two different customer tables, and you want to maintain a consistent standard for these codes. Cross-domain analysis enables you to directly compare these code values.

Testing—Mostly about Verification

Program testing can be used to show the presence of bugs, but never to show their absence.

—Dr. Edsger W. Dijkstra

Most complex software quality assurance activities are centered around testing. The general aim of testing is to affirm the quality of software systems by systematically exercising the software in carefully controlled circumstances. A good test is one that has a high probability of finding an as yet undiscovered defect. A successful test is one that uncovers an undiscovered defect.

Software testing maturity model is based on systematic CMM (Capability Maturity Model) methodology for continuous process improvement. Each testing maturity level has its set of processes and practices and are directly related to CMM laves. The purpose is to provide testing framework for assessing the maturity of the test processes in an organization, providing targets for improving testing maturity.

This testing strategy defines five levels of software testing maturity, as follows:

Level 1—Initial: At this level, an organization is using ad-hoc methods for testing, so results are not repeatable, and there is no quality standard. This looks more like debugging than testing. The objective of this kind of testing is to prove that the application has no defects.

Level 2—Definition: At this level, testing is defined as a process, so there might be test strategies, test plans, and test cases, based on requirements. Testing does not start until products are completed, so the aim of testing is to compare products against requirements.

Level 3—Integration: At this level, testing is integrated into a software life cycle, e.g., the V-model. The need for testing is based on risk management, and the testing is carried out with some independence from the development area.

Level 4—Management and Measurement: At this level, testing activities take place at all stages of the life cycle, including reviews of requirements and designs. Quality criteria are agreed upon for all products of an organization (internal and external).

Level 5—Optimization: At this level, the testing process itself is tested and improved with each iteration. This is typically achieved with tool support, and also introduces aims such as defect prevention throughout the life cycle, rather than defect detection (zero defects)[8].

Testing has been widely used as a way to help software engineers develop high-quality systems. The maturation process through which the testing techniques for software have evolved, from an ad hoc, intuitive process to an organized systematic software engineering discipline.

Part of a strategy document is to identify the type of testing required. The primary focus in this section is to describe each test identified in the testing strategy with the focus on the verification that correct data is transferred into the data warehouse.

Limitation of testing: Any nontrivial system will have defects. We cannot know in advance which inputs and in which sequence of execution will cause system to fail. Outcome may depend on internal state of the system when a particular code is executed. If a particular function returns a correct result, we cannot guarantee that function will always return correct results.

And it is impossible to test a function for all possible vales. Testing is not a substitute for defect prevention software engineering practices. Testing should be complemented by other (inspection, for example) software engineering practices.

---

[8]   CrossTalk magazine Aug. 1996: Ilene Burnstein, Taratip Suwannasart and C.R. Carlson: Developing a Testing Maturity Model

Unit Testing during Data Loading

A test that reveals a bug has succeeded, not failed.

—Dr. Boris Beizer, *Software Testing Techniques*

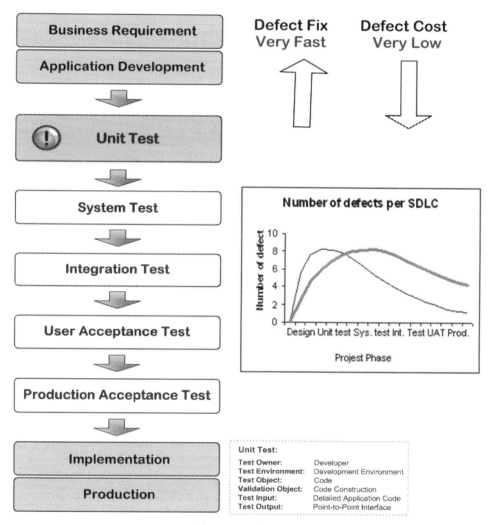

Fig 7.1 Unit Test

A unit test starts after the design reviews and code inspections are completed. Depending on the outcome of these pre-unit test activities, a project defect trend can take one of the two lines on the Number of Defects per SDLC graph on the previous page.

If the design inspections and code reviews are done correctly, as prescribed by CMMI standards, the number of problem potential defects will be uncovered early.

In fact, some of the most difficult, i.e., structural problem with the code, can only be uncovered during the design review or code inspection stage.

It is sometimes difficult or even impossible to create conditions in test environment to uncover these kinds of problems. If the defects are not found out during unit testing, we will be finding and fixing them in system testing, or worse, in the field, where the cost will be orders of magnitude higher. But rest assured that in the dynamic of the production environment, they will show up sooner rather than later. The testing strategy must take these important activities into consideration and strictly enforce them as they have a serious impact on quality of the delivered product.

Unit testing is done at the lowest level. It tests the basic unit of software, which is the smallest testable piece of software, and is often called *unit*, *module*, or *component* interchangeably.

Unit testing is the lowest level (i.e., a unit is a piece of source code) using a coverage tool; as a minimum, sufficient testing to assure that every source statement has been executed at least once. It may be necessary to test each branch (both *true* and *false*) to assure the 100 percent branch coverage of a unit of code.

Unit testing is done to find unit bugs, which is the second-highest frequency kind of bug. Unit testing refers to the testing of discrete program modules and scripts. This has traditionally and logically been the task of the developer.

Every developer must test her program modules and scripts individually. This type of testing is known as a white-box testing. The purpose of this test is to ensure the module or unit of code is coded as per agreed upon design specifications. The developer should focus on the following:

1. That inbound and outbound directory structures are created properly with appropriate permissions and sufficient disk space. All tables used during the ETL are present with necessary privileges.
2. That the ETL routines give expected results.
3. Data cleansing which includes the following:
    -   ETL transformation logics from source to target work as specified in designed documentation.
    -   Boundary conditions are satisfied (e.g., check for date fields with month of February, leap year dates, zero or less years of age, etc.).
    -   Surrogate keys have been created as primary key as applicable. (Conversely, neither natural nor compost keys should not be used as primary key). Database indexes and referential integrity constraints can help avoid some these problems, but not all data is within a relational

database. Some of the relations cross databases and technologies, but even within databases, not all data can afford to implement referential integrity constraints.

- Appropriate fields are initiated with NULL values where expected.
- Log of rejects are created where applicable, with sufficient details.
- All error recovery methods and/or routines are verified. Sometimes, during testing, it is not possible to create condition, which would invoke an error recovery method. The code involved in this particular situation must be reviewed with peers or appropriate knowledgeable team members.
- Auditing for any update, delete or insert into DWH is done properly.

4. That the data loaded into the target is complete:
   - All source data that is expected to get loaded into target actually got loaded.
   - Pre and post row counts and checksums work wonders here.
   - All fields are loaded with full content (i.e., no data field is truncated while transforming).
   - No duplicates are loaded.
   - Aggregations take place in the target properly.
   - Data integrity constraints are properly taken care.

Unit Test Automation

An automated unit test is a piece of code written by a programmer that exercises another piece of functional code. Unit testing is performed to verify that a piece of code does what the programmer thinks it should do. Unit testing is relatively easy practice to adapt, but there is a set of guidelines that has to be followed to make it affective. This is demonstrated here:

There exists an open source unit testing framework, initially develop by Steven Feuerstein, modeled after Junit and xUnit frameworks. Commercial version is available from Quest and Steven Feuerstein.

Here, the demonstration of the steps involved in unit testing Pl/SQL procedure:

1. Select the table and PL/SQL procedure to test.
2. Create the Unit Testing Repository.
3. Create the Unit Test.
4. Run Unit Test
5. Create and run and run an Exception Unit Test
6. Create a Unit Test Suite
7. Run Unit Test Suite and produce the report.

Unit Test Automation Example

The Quest tool Code Tester for Oracle is used to demonstrate unit test process automation. In this example, VENDOR_RATING field has become mandatory for all active accounts. If the VENDOR_RATING value for an active member is missing, a default value of 100 is inserted. For the vendors where there is a value in the field VENDOR_RATING, the field is not updated.

Create the VENDOR_INFO Table with sample data:

```
CREATE TABLE vendor_info (vendor_id NUMBER PRIMARY KEY, vendor_name
VARCHAR2(30),  active_flag CHAR(1), sale_amt NUMBER, vendor_rating
NUMBER, modified_date DATE);

INSERT INTO vendor_info VALUES (10002001, 'ABC Company', 'Y', 98000, 101,
'01-jan-10');

INSERT INTO vendor_info VALUES (10002003, 'TTT INC', 'Y', 200000, null,
'01-jan-10');

INSERT INTO vendor_info VALUES (10002021, 'DDD TECH', 'Y', 1000, 91,
'01-jan-10');

INSERT INTO vendor_info VALUES (10002035, 'CC INC', 'Y', 0, null,
'01-jan-10');

INSERT INTO vendor_info VALUES (10002061, 'R BANK', 'N', 5000, null,
'01-jan-10');

INSERT INTO vendor_info VALUES (10002062, 'B INC', 'Y', 39000, 80,
'01-jan-10');

INSERT INTO vendor_info VALUES (10002094, 'D Computer INC', 'Y', 2000,
null, '01-jan-10');

INSERT INTO vendor_info VALUES (10002097, 'Tom INC', 'N', 2000, 70,
'01-jan-10');
```

VENDOR_ID	VENDOR_NAME	ACTIVE_FLAG	SALE_AMT	VENDOR_RATING	MODIFIED_DATE
10002001	ABC Company	Y	98000	101	1/1/2010
10002003	TTT INC	Y	200000		1/1/2010
10002021	DDD TECH	Y	1000	91	1/1/2010
10002035	CC INC	Y	0		1/1/2010
10002061	R BANK	N	5000		1/1/2010
10002062	B INC	Y	39000	80	1/1/2010
10002094	D Computer INC	Y	2000		1/1/2010
10002097	Tom INC	N	2000	70	1/1/2010

Fig. 7.2

Create the RATE_UPDATE Procedure:

```
CREATE OR REPLACE PROCEDURE rate_update (ven_id NUMBER)
IS
    v_rate    REAL;
    v_active CHAR (1);
BEGIN
    SELECT vendor_rating
      INTO v_rate
      FROM vendor_info
     WHERE vendor_id = ven_id;
    SELECT UPPER (active_flag)
      INTO v_active
      FROM vendor_info
     WHERE vendor_id = ven_id;
    IF v_rate IS NULL AND v_active = 'Y'
    THEN
        UPDATE vendor_info
           SET vendor_rating = 100
         WHERE vendor_id = ven_id;
    END IF;
END;
```

Screen-print of the stored procedure in the TOAD® [Quest Software]:

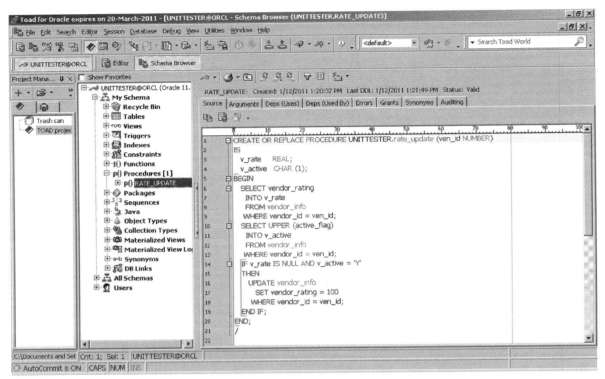

Fig. 7.3

```
CREATE OR REPLACE PROCEDURE UNITTESTER.rate_update (ven_id NUMBER)
IS
  v_rate    REAL;
  v_active  CHAR (1);
BEGIN
  SELECT vendor_rating
   INTO v_rate
   FROM vendor_info
  WHERE vendor_id = ven_id;
  SELECT UPPER (active_flag)
   INTO v_active
   FROM vendor_info
  WHERE vendor_id = ven_id;
  IF v_rate IS NULL AND v_active = 'Y'
  THEN
    UPDATE vendor_info
      SET vendor_rating = 100
    WHERE vendor_id = ven_id;
  END IF;
END;
/
```

Fig 7.4

Unit Testing Phase:

Create Test Matrix:

Test Case #	Expected Results	Test Data	Vendor ID
1	Insert the default vender rate	Active account VENDOR_RATING Null	'10002003'
2	The record is not be updated	Active account VENDOR_RATING not Null	'10002001'
3	The record is not be updated	Nonactive account VENDOR_RATING Null	'10002061'
4	The record is not be updated	Nonactive account VENDOR_RATING Not Null	'10002097'

Fig 7.5

Test Sample Tool: Quest Code Tester for Oracle—Beta Version 2.0:

Fig. 7.6

Step 1: Login to the Quest Code Tester for Oracle:

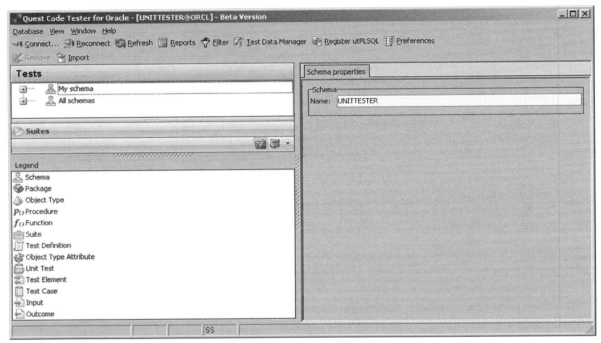

Fig. 7.7

Step 2: Choose the test procedure.

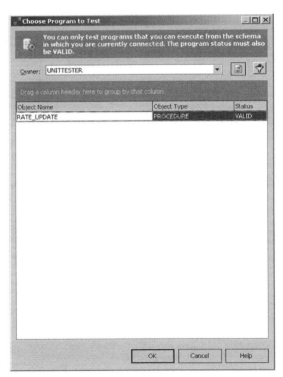

Fig. 7.8

Step 3: Insert the test cases.

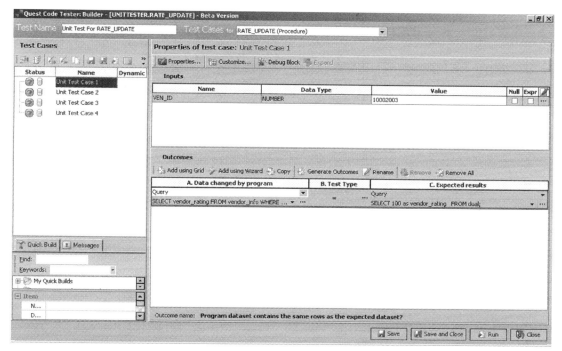

Fig 7.9

For example, Test Case 1 – Vendor ID = '10002003':

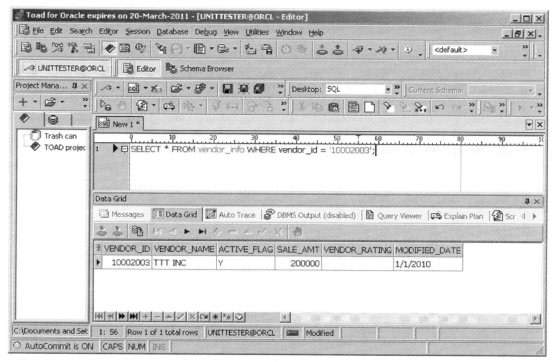

Fig 7.10

The ACTIV_FLAG value is set to *Y* and the VENDOR_RATING is set to *null*. This according to the business requirement is the condition that needs to be tested. This record should be updated after the procedure applied. Automated test must verify that this process works.

In the example above for VENDOR_ID = 10002003 is the input value in the Test Builder of Quest Code Tester. The outcome value for the field VENDOR_RATING for this record should populate a default value 100 after the procedure has ran.

SELECT vendor_rating FROM vendor_info WHERE vendor_id = '10002003';

Also make sure the related properties have been fully configured. In this case, for example, rollback is required after the execution.

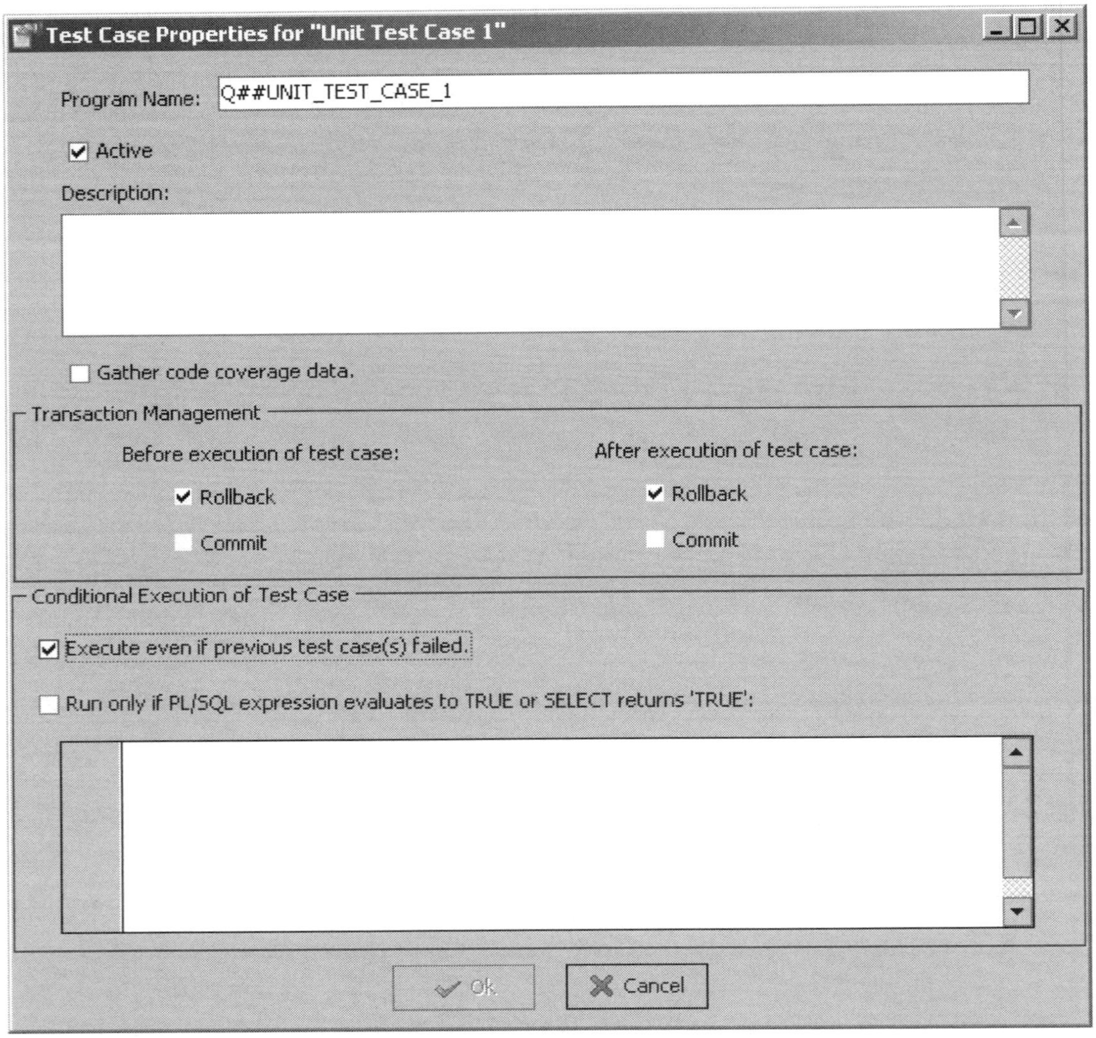

Fig 7.11

Step 4: Run all the test cases in Quest Code Tester.

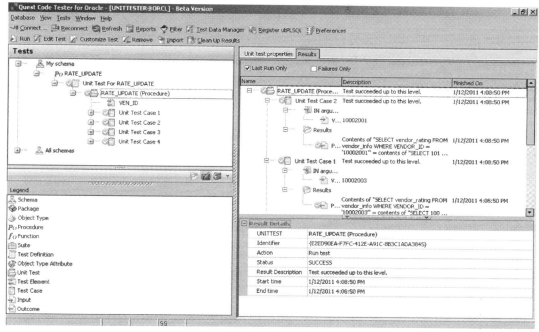

Fig 7.12

Step 5: Create Test Report.

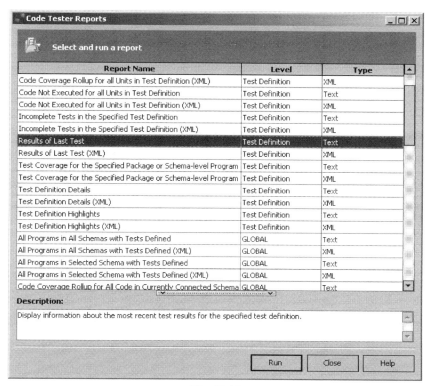

Fig. 7.13

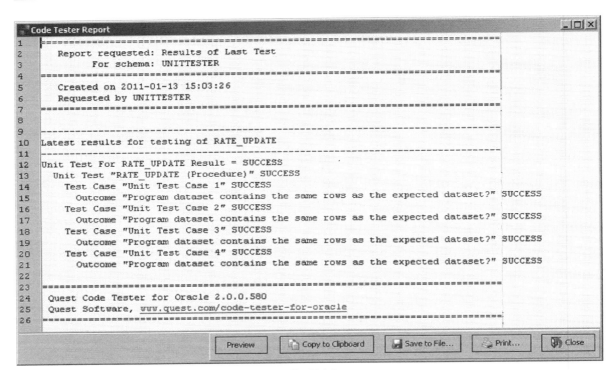

Fig 7.14

This process allows the code developer to create and run the atomic test case during code development, and at the same time accumulating full Test Suite in the Code Tester for Oracle tool. At the end of development, the full Test Suite is executed to confirm that the code is defect-free, before being delivered it to the next project phase.

Testing Stored Procedures

Introduction

Today's variety of data environments are increasingly challenging for test engineers. Applications, whether web-based or client-server, must be tested for seamless interface with the backend databases; this typically goes far beyond what the popular test automation tools can provide. Testers are required to know how to create and use SQL, stored procedures, and other database objects to effectively test today's data driven environments. They also need to know how to successfully test objects such as stored procedures and views in an application's databases.

What we hope you will take away from this section are the following:

- Why stored procedures are important to database testers?
- How to test an application's stored procedures in the database?
- Why testing stored procedures is necessary and why popular automated tools can't keep up?

This topic covers tips and techniques for creating and testing efficient stored procedures for database backend testing.

With the increasing use of data on the Internet and all the associated headaches, the need for skilled testers in this area is growing quickly. Stored procedures are increasingly used by experienced database developers to ease the complexity of calls to the database backend. These procedures contain critical tasks such as inserting customer records and sales. They need to be tested at several levels. Black box testing of the front-end application is important, but that makes it difficult to isolate the problem. Testing at the backend can increase the robustness of the data.

What to look for when testing stored procedures:

- Basic functionality; test input and output parameters using standard techniques (boundary analysis, parameter validation, etc.).

- Functionality for stored procedures that use triggers.
- Stored procedures which include queries that cover the entire table, i.e., table scans (performance).
- Stored procedures which output nothing.
- System/application errors returned to the user.
- Corrupt data as output.
- Susceptibility to deliberate, destructive attacks such as SQL injection attacks.

It's possible to use scripts to submit to the RDBMS that will test a variety of inputs to the critical backend stored procedures. For example, a stored procedure to insert data into the database can be tested by using a script that calls that procedure. Batch files or a table of test data can be used as the automated input to this routine. Similar scripts can be used to form a test bed for the critical database procedures. Once the test is completed and determined to be useful, it can be scheduled to run again at a future time as a batch. Major RDBMS have scheduling capability as well—in SQL Server, this is called a job.

Individual stored procedure tests

Verify the following and compare them with design specification.

- Whether each stored procedure is installed in its assigned database.
- Stored procedure name vs. standard naming conventions.
- Parameter names, parameter types and the number of parameters.

Stored procedure outputs

Verify
   ◦ When output is zero (zero rows affected).
   ◦ When some records are extracted and others are not.
   ◦ What each stored procedure is supposed to do.
   ◦ What each stored procedure is not supposed to do.
   ◦ With simple queries to see whether each stored procedure populates correct data.

Stored procedure parameters

Verify
   ◦ Check parameters that are required.
   ◦ Call stored procedures with valid and invalid data.
   ◦ Call procedures with boundary data.
   ◦ Make each parameter invalid and run each procedure.

Stored procedure return values

- Whether a stored procedure returns values and if values are correct.
- When a failure occurs, nonzero must be returned.
- Stored procedure error messages
- Make stored procedure fail and cause every error message to occur at least once.
- Find out any exception that doesn't have a predefined error message.

Others:

- Whether a stored procedure grants correct access privilege to a group/ user.
- See if a stored procedure hits any trigger error, index error, and rule error.
- Look into a procedure code and make sure major branches are test covered.

Integration tests of stored procedures

- Group related stored procedures together. Call them in particular order.
- If there are many sequences to call a group of procedures, find out equivalent classes and run tests to cover every class.
- Make invalid calling sequence and run a group of stored procedures.

Design several test sequences in which end users are likely to do business and do stress tests.

Updating triggers

Verify the following and compare them with design specification. Make sure trigger name is spelled correctly.

- Determine whether a trigger is generated for a specific table column and verify the trigger's intended action.
- Update a record with valid data.
- Update a record that a trigger prevents with invalid data and cover every trigger error.
- Update a record when it is still referenced by a row in another table.
- Verify the rolling back of transactions when a failure occurs.
- Find out any case in which a trigger is not supposed to roll back transactions.

Inserting triggers

Verify the following and compare them with design specification.

- See if a trigger is generated for a specific table column.
- Verify trigger's insertion function.
- Insert a record with a valid data.
- Insert a record a trigger prevents with invalid data and cover every trigger error.
- Try to insert a record that already exists in a table.
- Make sure rolling back transactions works when an insertion failure occurs.
- Identify any case in which a trigger should roll back transactions.
- Identify any failure in which a trigger should not roll back transactions.
- Identify conflicts between a trigger and a stored procedure/rules (i.e., a column allows NULL while a trigger doesn't).

Deleting triggers

Verify the following and compare them with design specification.

- Make sure trigger name spelling is correct.
- See if a trigger is generated for a specific table column.
- Delete records to verify trigger deletions.
- Delete a record when it is still referenced by a row in another table.
- Verify every trigger error.
- Try to delete a record that does not exists in a table.
- Make sure rolling back transactions when a deletion fails.
- Identify cases where a trigger should roll back transactions.
- Identify failures in which a trigger should not roll back transactions.
- Conflicts between a trigger and a stored procedure/rules (i.e., a column allows NULL while a trigger doesn't).

System Testing

Testers are people who realize that things can be different!

—Jerry Weinberg

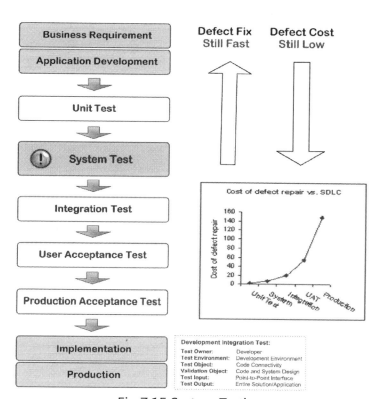

Fig 7.15 System Testing

Typically, the scope of system testing only includes testing within the ETL application. The endpoints for system testing are the input and output of the ETL code being tested. The validated design document is the Bible, and the entire set of test cases is directly based upon it. The functionality of the application is tested; usually, it is done using the black-box test methodology. The major challenge here is the preparation of intelligently designed input test dataset. This can uncover most of the problems in the ETL application quickly.

Production-like data should be used wherever possible. We must select a test data to verify for all possible combinations of input and specifically check for errors and exceptions. Knowledge of the business process and business rule for data transformations are essential here, as we must be able to relate the results functionally and not just verify the code.

The testing team must verify for:

Data completeness, or conversely, no *data losses*—Includes validating that all records, all fields and the full content of each field are loaded. Strategies to consider include:

- Record counts between source data, data loaded to the warehouse and rejected records should be reconciled.
- Negative scenarios are also validated (Verifying for special characters, etc.).

Correct data transformation—All data is transformed correctly in accordance with business rules as per design specifications. This could be simple "one to one" or complex data transformation.

Data aggregations—Match aggregated data against staging tables and/or ODS.

Data quality—The ETL application correctly rejects and substitute default values, corrects or ignores and reports on invalid data. ETL testing revolves around the data, that is why is important to achieve a degree of excellence in data accuracy. We know we have achieved quality when we successfully fulfill customers' requirements. Data accuracy will deliver a value for our customer.

We will discuss this subject in further detail in the *Integration Testing* section.

Data Completeness

The most basic test of data completeness must be executed first. Objective of this test is to verify that all the expected data has been loaded into the data warehouse. This includes validating that all records, all fields and the full contents of each field are loaded. Testing scenario to consider in this verification includes:

- Verifying the count of record in the source data and data loaded to the warehouse plus the rejected records are equal.
- Comparing the unique values of key fields between source data and destination data (data loaded in the warehouse). This technique may points out a variety of possible data errors without having to do full validation on all fields.
- Utilizing a data profiling tool or SQL filters that show the range and value distributions of fields in a data set. This can be used during testing and in

production to compare source and target data sets and point out any data anomalies from source systems that may be overlooked if the data movement is correct.

- Populating the full content of each field to validate the contents of each field ensuring that no truncation occurs at any step in the process. For example, if the source data field is a string 30, then make sure to test it with 30 characters.
- Testing the boundaries of each field to find any database limitations. For example, if a numeric field is defined as an integer whose length three (2) decimal number, it must include all values between 9 and 99. If the field is defined date fields, it must include the entire range of valid dates.

Log of bad records should be kept in a log table for audit and continues process improvement purposes. This concept should be carried over into production environment.

Data Transformation

Verifying that data is transformed correctly based on business rules can be the most complex part of testing an ETL application with significant transformation logic. Data transformation is mostly the application of business rules that need to be applied to the data. In data warehousing, this may include the computation or derivation of new values from existing data (such as profitability from cost and revenue), normalization or denormalization, filtering, classification, aggregation, or summarization of the data, etc.

One typical method of verification is to pick some sample records and compare by eyeballing to first validate data transformations manually. This is used as initial test step to do rudimentary verification, but it is a manual testing and requires full understanding the ETL logic. A combination of automated data profiling and automated data movement verification is a better long-term strategy.

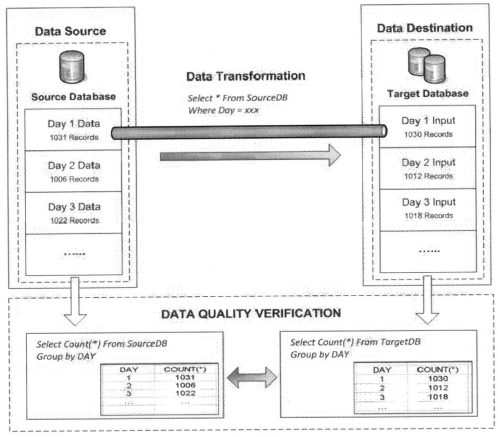

Fig 7.16

Here are some simple test scenarios that can be easily employed to validate and verify data movement within the DWH:

- Create sets of scenarios of input data and expected results and validate these with the business customer. This is a good requirements elicitation exercise during design and can also be used during testing.
- Create test data for all the above scenarios. Elicit the help of an ETL developer to automate the process of populating data sets with the scenarios all possible data, making them test driven scenarios.
- Utilize data profiling results to compare the range and distribution of values in each field between source and target data.
- Verify correct processing of ETL-generated fields such as default values or surrogate keys.
- Verify that data types in the warehouse are the same as specified in the design and the data model.
- Scenarios that test referential integrity between tables. For example, what happens when the data contains foreign key values that are not in the parent table? Test how orphaned child records are handled.

Enterprise Integration Testing of the DWH

To find the bugs that customers see—that are important to customers—you need to write tests that cross functional areas by mimicking typical user tasks. This type of testing is called scenario testing, task-based testing, or use-case testing.

—Brian Marick

The purpose of integration testing is to show how the DWH application works from an end-to-end perspective. We must consider the compatibility of the DWH application with all upstream and downstream flows, assuring data integrity across the flow. Following our notion of Validation and Verification, the first level of integration testing begins with validation of the warehouse data model. All subsequent testing is based on understanding this model and indeed, each element in the model and how it contributes to achieving specific business objectives in the BI area.

Validation

Granularity refers to the level of detail of the data stored in fact tables in a data warehouse. High granularity refers to data that is at or near the transaction level. Data that is at the transaction level is usually referred to as atomic level data. Low granularity refers to data that is summarized or aggregated, usually from the atomic level data. Summarized data can be lightly summarized data, as in daily or weekly summaries, or highly summarized data, such as yearly averages and totals.

Granularity is the single most important aspect to consider when designing a data warehouse. Data should be captured at its lowest, most atomic grain. Atomic data is highly dimensional. It is always possible to create higher level grain by aggregating atomic level data for a business process that may require it. The more detail there is in the fact table, the higher its granularity, and vice versa. The consequence of the higher granularity of a fact table is that more rows will be required to store it. Let us look at an example to illustrate this important point:

We are going to take a look at a small data warehouse with a single fact (Sales) and three dimensions (Time, Organization, and Product). The fact table contains three metrics (Unit Price, Units Sold, and Total Sale Amount). The time dimension consists of four hierarchical elements (Year, Quarter, Month, and Day). The organization dimension consists of three hierarchical elements (Region, District, and Store). The product dimension consists of two hierarchical elements (Product Family and SKU).

The metrics in the Sales fact table must be stored at some intersection of the dimensions of Time, Organization, and Product. Therefore, in this data warehouse, the highest granularity that we can store Sales metrics in is in Day/Store/SKU (as these are the lowest level in each dimensional hierarchy). Conversely, the lowest granularity that we can aggregate Sales metrics to in this data warehouse is by Year/ Region/Product Family (as those are the highest levels in each dimensional hierarchy). We may also (for a variety of performance reasons) choose to store Sales metrics at some intermediate level of granularity, such as by Month/District/SKU. In other words, summarized data can be lightly summarized, as in daily or weekly summaries, or highly summarized, as in yearly averages and totals.

Assuming that the star schema discussed in the above paragraph of 500 products, sold in 200 stores, and being tracked for 10 years, with the daily grain. This could produce a maximum of 365,000,000 records in the fact table. This is a design parameter that is used in determining the size for the data warehouse. Moving from the daily to a weekly grain in the same example, would reduce the maximum number of records to somewhat more than 52,000,000.

The question we need to answer is: which level of granularity is right? The answer to this question depends on what business question this data warehouse needs to answer. Quite often, we choose to have fact tables with a high level and low (aggregated) level of granularity. That is why the question of right granularity for a data warehouse is an important issue to ponder upon when designing DWH.

Storing data at the highest level of granularity appears not to be a bad idea. The data can be reshaped to meet different needs of the finance department, of the marketing department, of the sales department, and so forth. Granular data can be summarized, aggregated, broken down into many different subsets, and so forth. There are indeed many good reasons for storing data in the data warehouse at the lowest level of granularity. Many companies start out by doing just that, building a data warehouse at the highest level of granularity, and find that their true requirements and budget make building this Enterprise DWH model impractical. Only at that time do they start evaluating how the system is going to be used. What kind of reporting will be generated off it? It may also not be a bad idea to evaluate a business vision and the users' expectations. Take all of this into account and build the database for efficiency too.

The impact of the granularity of a data warehouse is profound, and it has to be carefully evaluated as it affects the size and the performance of the DWH. At the same time, use of DWH in the future has to be taken into account, as the implemented granularity is hard to change, but it may impede the needs of business in the future.

Changing Attributes

One of greatest challenges in data warehouse is the problem of changing attributes. Let us take an example from the sales schema discussed above. In the store dimension each store is located in a particular sales region, territories and the zone. Occasionally, to reflect changing business condition, some realignment of the stores and regions may be required. If the organization simply updated the table with new realignment, and if the user tries to look for at historical sales for the region, the number will not be accurate. The total sales will reflect the current structure of the star schema. In other words, the business has lost true history. Sometimes, this may be intentional business decision, but more often it not.

Typical solution to this problem is to create the new record for the store. The new record containing the new realignment, but leave the old record with the old realignment intact. This approach prevents from comparing the same store current sales to its historical sales as the StoreID (primary key) must be different for new and old record. The solution to this problem is to use StoreName (as it is the same on both records) together with BeginDate and EndDate fields.

Potential Problems with Fact Tables

The purpose of the fact table is to be able to easily get useful information out of the data. In order to do this, any complex math or unusual query requirements should be avoided. The easiest way to do this is to make sure the actual measures, the numbers, are additive across all rows.

The additive property of the numbers across the rows is a most precious gift that must always be preserved in EDW. An accurate summary of data is of paramount importance in OLAP. The DWH is conceptualized around facts and dimensions tables. Facts contain a measure of interest and dimensions have attributes used to select and aggregate measure of interest. Classification attributes are modeled in forms of hierarchies.

For example, a classification of hierarchy of time detention shows that measure in the fact table can be aggregated from the lowest level of granularity—seconds—to progressively higher levels—hours and days. As an example, the density of telephone traffic may be aggregated from density of traffic per second to daily density of telephone traffic. Aggregation is usually done along multiple dimensions. We may want to know the daily density of traffic in Toronto for 2008. Note that the date dimension may be a separate dimension. It is important to understand that measure can be additive across all dimensions, one dimension, or not additive across any dimension. The first step in ensuring accuracy of aggregation is to recognize when and how aggregation operation can be used.

Some examples of *nonadditive measure* are as follows:

- *Ratios*—e.g., ratios of men to women in different departments of a company
- *Measure of intensity*—e.g., temperature, speed
- *Percentages*—e.g., return on investment percent
- *Maximum*—e.g., temperature, blood pressure
- *Minimum*—e.g., income, account balance
- *Averages*—e.g., account balance, arrival time
- *Code* (used for identification)—e.g., social insurance number (SIN), postal code, barcode
- *Sequential numbers*—e.g., order number, ID number

Examples of *semiadditive measure*:

- *Changing data*—e.g., address, telephone number
- *Dirty Data*—e.g., duplicated data, incorrect data

Averages, maximums, minimums and so on are a big problem, because they appear to be regular numbers but they are definitely not additive. Summing averages, or percentages, or ratios, or anything like that gives useless results that do not appear to be useless.

Let's say, for example, that we need to make a fact table with sales volumes reported by region, sales team, product, date, product supplier, etc. We decide that we need daily totals and monthly totals, but we do not want to calculate the monthly numbers on the fly all the time, and since both will be using the same dimensions, why not put them in the same fact table?

If we do this, then we can put an indicator in the fact table row, showing some records as daily and some as monthly, or we can just put the previous months' information duplicated for each day in an extra column in the daily rows. There are most likely other ways of doing this. The problem we have now is:

One, we need to remember to use *Select Distinct* to get the monthly data, which increases the complexity of the queries, and we need to make sure everyone who ever uses this table for as long as it is in existence knows about the requirement to use *Select Distinct*.

Two, what if there is a problem? What if one of the days was reported wrong? Then we need to either update the rows in the fact table a (big no, no) or mark the rows as deleted and insert a bunch of new rows. Another problem—what if one of the dimensions changes part-way through the month? The daily data would still be valid, but the monthly totals would need to change part-way through.

A good solution to this problem would be to just use two fact tables, or even better, do the aggregation on the fly. These are just some of the issues that could be caused by mixed granularity.

In the past, applications were developed independently without any consideration that the data that were used in the applications would ever have to be integrated with other applications' data into the corporate data image. However, one important aspect of the data warehouse is the integration of all corporate data from multiple and very often incongruent data sources.

As the data from online transaction processing (OLTP) systems, batch systems, and from externally syndicated data sources is fed into DWH, it is being transformed and reformatted into one single corporate image for each entity. Because data from two different systems is to be merged together in the warehouse, it is obvious that we have to reconcile the different representations of the same data.

This leads to an unintended benefit of an enterprise-wide process of data standardization and had been talked about for some time as a *desirable goal*, but no one has ever been given the responsibility of spearheading this effort. Drawing attention to these disparities is an eye-opener to many who have had no idea that other units of the enterprise were representing identical information quite differently. The diagram below illustrates this aspect of data integration as it passes from the operational environments or the external syndicated feeds into to the EDW.

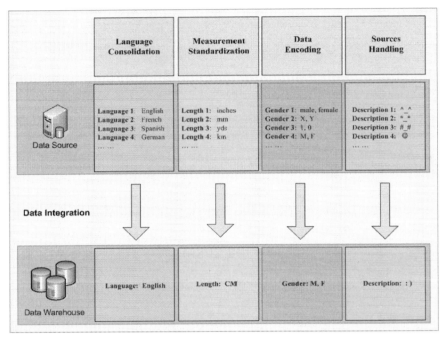

Fig. 7.17 Data integration

- Table names should be intuitive, so a formal definition of the table may not be required but is nonetheless still useful.
- A table is best described by the data elements that reside within it. Each table should be identified by how it is to be used within the warehouse, e.g., fact table, dimension table, or summary table.
- Primary keys should be identified and placed at the top of each table for quick reference.
- Validation that the primary keys in the dimension tables are actually unique identifies each row in a table.
- It should be validated that the primary key in a dimension table uniquely identifies each row in the table.
- It is important that the primary keys of dimension tables remain stable. Therefore it is strongly recommended that the surrogate key, also known as an artificial or identity key, a substitution for the natural primary key, be created and used for primary keys for all dimension tables. The only requirement for a surrogate primary key is that it uniquely identifies each row in the table.

The surrogate key is useful because the natural primary key—e.g., the customer number in the customer table or the city name in the address table—may change.

Source data tables may use different keys for the same entity. Legacy systems that are provided might have used a different numbering system than a current online transaction processing system. The surrogate key uniquely identifies each entity in the dimension table regardless of its source key. Slowly changing the dimension can only be implemented using a surrogate key. Dimension tables should be similar across all the fact tables.

Verification

The second part of the test strategy should include testing or verification for:

1. *Sequence of jobs*—to be executed with job dependencies and scheduling
2. *Notifications of problems issued*—given to IT and generated in proper format
3. *Accuracy*—of the results.
4. *Restartability of jobs*—in case of failure.
5. *Generation of error logs and audit tables*—generated and populated properly.
6. *Cleanup scripts*—for the environment including the database.
7. *Granularity*—of data is as per design specifications.

The combined responsibility for and participation in this activity by experts from all related applications is a must in order to avoid misinterpretations of the results.

When creating integration test scenarios, consider how the overall end-to-end process works and can break. Focus should be on intermediate points (points between the applications) between applications, not on any application involved. Testing should include consideration of process failures at each step, how the flexure is handled and how data would be recovered or deleted if necessary. The intermediate points (interfaces) are the areas where the defects are found.

Most defects found during integration testing are found, surprisingly, at interfaces. Defects are either data related or resulting from false assumptions or misunderstandings about the design of another application. Therefore, it is important to do the integration test with production-like data.

Real production data is ideal, but depending on the content of the data, there could be privacy or security concerns that require certain fields to be randomized before using it in a test environment. Communication between all groups—QA, BA, and Development—is a must. To bridge the communication gap, team members from all systems are asked to review and formulate test scenarios and discuss what could go wrong in production. There should be a dry run of the overall process from end to end in the same order and with the same dependencies as in production. Integration testing should be a combined effort and not solely the responsibility of the QA team.

Regression Testing

This is really, the equivalency testing, in which we rerun the same suite of tests to assure that the current version has not inadvertently regressed to the previous version especially in areas which have been changed. During the test cycle, anytime actual test results do not match the expected test results the program producing the error must be fixed, and all programs must be rerun.

A DWH application is not a one-time solution. Possibly, it is the best example of an incremental design where requirements are enhanced and refined quite often based on business needs and feedbacks. In such a situation, it is very critical to the test that the existing functionalities of a DWH application are not messed up whenever an enhancement is made to it. Generally, this is done by running all functional tests for existing code whenever a new piece of code is introduced. However, a better strategy could be to preserve earlier test input data and result sets and run them again. At that point the new results could be compared against the older ones to ensure proper functionality.

Regression testing is a revalidation of existing functionality with each new version of code. Due to defect fixing and/or enhancements in the DWH application, the regression test suite will be executed multiple times. That is why the regression test suite is a good candidate for automation, start building regression/automation test suite during system testing. Regression testing process runs much smoother if the regression test suite has been automated. Test cases should be prioritized by risk in order to help determine which ones need to be rerun for each new release. A simple but effective and efficient strategy to retest basic functionality is to store source data sets and results from successful runs of the code and compare new test results with the previous runs.

When doing a regression test, it is much quicker to compare results to a previous execution than to do entire data validation again. The recursion test, or any test that needs to be repeated, should be automate.

Performance Testing

In addition to the functional tests described above, a DWH may need to go through another phase called performance testing. Any DWH application must be designed to be scalable and robust. This must be verified before DWH application goes to production environment. The performance concerns are in two primary areas:

1. The speed for data cleansing and staging after an accounting period has closed.
2. Performance of data retrievals for the information consumers.

Performance test must be done with a large production-like volume of data. Objective is to ensure that the load window is met with production-like volumes. This phase should involve the DBA and ETL teams and others who may be able to review and validate the code for performance optimization.

The primary purpose of a DWH application is to ensure that the data is accurate and credible. While this is necessary, it is not enough. Where DWH application typically falls short is in delivering information value. Data is not information. The correct formula for information is relevance of data in that context. Providing information from the DWH application into consumers' hands as they need it is a key part of providing information value. Data cleansing and staging may take a long time, and by that time, the information may lose its value relevance.

As the volume of data in a data warehouse grows, the ETL load times can be expected to increase and performance of queries can be expected to degrade. Having a solid technical architecture and a good ETL design can mitigate this. The aim of the performance testing is to point out any potential weaknesses in the ETL design, such as reading a file multiple times or creating unnecessary intermediate files. The following testing scenarios would help discover performance issues:

- Load the database with peak expected production volumes to ensure that the ETL process within the agreed-upon window can load this volume of data.
- Compare these ETL loading times to loads performed with a smaller amount of data to anticipate scalability issues. Compare the ETL processing times component by component to point out any bottlenecks (areas of weakness).
- Monitor the timing of the reject process and consider how large volumes of rejected data would affect the performance.
- Perform simple and multiple join queries to validate query performance on large database volumes. Work with business users to develop sample queries and acceptable performance criteria for each query.

Materialized views in DWH are built with performance improvement in mind. Unlike an ordinary view, which does not take up any storage space because they are generated on the fly, materialized views provide indirect access to table data by storing the results of an aggregation query in a separate schema object. A materialized view definition can include any number of aggregations: [SUM, COUNT(x), COUNT(*), COUNT(DISTINCT x), AVG, VARIANCE, STDDEV, MIN, and MAX]. Materialized summary tables use *aggregate keys* to define a hierarchy of aggregation. Several dimensions can be contained in each aggregate key.

Performance testing for retrievals at different level of aggregation is required to verify acceptable performance at all levels of *aggregations*?

Materialized views can be stored as summary tables in the same database as their base tables. This can improve query performance within OLTP systems. Most databases do not have enough CPU capacity to handle both OLTP and the heavy demands of OLAP processing. To ensure good performance, materialized views are made available in a separate schema or even on another machine than the OLTP system. The two databases are synchronized on a nightly (rather than real-time) basis.

User Acceptance Testing [UAT]

The map is not the territory. When the map and the territory disagree, believe the territory.

—Jerry Weinberg

Acceptance Testing is testing from the users' perspective, typically end-to-end, to verify the operability of every feature. UAT is done after the completed system is handed over from the developers to the customers or users. The purpose of acceptance testing is rather to give confidence that the system is working than to find errors.

The raison d'être of a data warehouse application is to make data available to business users. Users know the data best, and their participation in the testing effort is essential to the success of a data warehouse. We build software for users. They're not (should not be) concerned with how the software works, how it's organized, or the cute programming tricks we used. They're only concerned with how it behaves and therefore, testing that behavior must be central to this testing effort. Therefore, acceptance testing must be performed by business users. They are the most knowledgeable to validate the functionality of the data warehouse application. This also helps in avoiding "Yes, this is nice, but it would be nicer if . . ." syndrome. Quality assurance assures and the development team supports the UAT team during this test. This is the most critical part of the test cycle because the actual users are the best judges to ensure that the application works as expected by them. Business users, for example, do not have ETL knowledge. That is why they must be supported during this test by quality assurance and development teams. The UAT test exit report is signed off by the users and they must be able to understand it before they sign-off on it.

User-acceptance testing typically focuses on data loaded into the data warehouse and any views that have been created on top of tables.

1. Production (if possible) or near-production data should be used for UAT. Users typically think of issues in terms of the *real* data.
2. Test database views are compared with expected results.

3. The QA test team must support users during UAT. This is because the end users will have questions about how the data is populated and need to understand details of how the ETL works.
4. During UAT, data will need to be loaded and refreshed a few times.

Regression Test Planning for the Data Warehouse

A few meanings of regression testing from the experts:

1. Testing that is performed after making a functional improvement or repair to a report or data. Its purpose is to determine if the change has regressed other aspects of the report or data.
2. A repetition of tests intended to show that the software's behavior is unchanged except as required by change to the software or data.
3. Testing conducted for the purpose of evaluating whether a change to the system has introduced a new failure.

The figure below shows stages of regression testing that might be necessary after changes to reports or data.

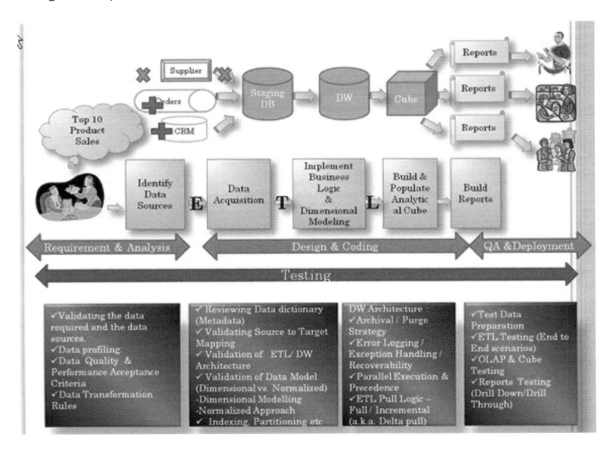

Common Strategies for Selecting Regression Test Suites

1. *Retest High Priority, Risky Use Cases/Test Cases* Choose baseline tests to rerun by risk heuristics—most risk to data, report or dashboard when failing.
2. *Retesting By Profile* Choose baseline tests to rerun by allocating time/QA resources in proportion to operational profile.
3. *Retesting Changed Segments* Choose baseline tests to rerun by assessing code or data changes.
4. *Retest Using Existing Test Cases* Choose baseline tests to rerun by analyzing dependencies and relationships with changed or added code.

A Suggested Strategy for Regression Testing Data

Recommended: Combine four common strategies from above.

Any one of the four regression testing strategies described earlier may be good, but in the real world a combination of the strategies may be a better decision.

It is assumed that first we test the data change (ex. ETL fix) itself by running all related test cases. Then for regression testing:

1. <u>30 percent of the allowed regression test time:</u> Run test cases representing the most risky functions, in particular, those that may be affected by new or changed code. Components for defining priority one consider business risk and frequency of using this scenario by the customer.
2. <u>50 percent of allowed regression test time:</u> Continuous cycle regression testing for running the existing regression test suite.
3. <u>20 percent of allowed regression test time:</u> Exploratory Testing—remember to properly document the results of exploratory testing. As a minimum, update the test log. If you do not like the sound of *exploratory testing*, use the time in the schedule to improve your understanding of the requirements, the system and your logical and architectural coverage of application by TC. You must allocate time and resources for this task anyway.

50 percent, 30 percent, and 20 percent are a common allotment; others make work better. You can play this around as you like. Most importantly, all your existing TC will run, you will have the priority for running TC, and you will allocate time for keeping your TC suit under continuous improvement.

The selection of test cases for regression testing . . .

- Requires knowledge of the bug fixes and how they affect the system.
- Includes areas of frequent defects.

- Includes areas which have undergone many and/or recent code changes.
- Includes areas which are highly visible to the users.
- Includes the core features of the data which are mandatory requirements of the customer.

What's Needed for ETL Regression Test Plans?

After a data load or data table change of any kind, the following questions should be pursued to plan for testing.

1. Which <u>ETL processes</u> were changed and what new or changed logic was implemented?
2. Which <u>stored procedures</u> and materialized views were changed; what were the changes to logic?
3. Which <u>tables, views</u> and related fields were changed?
4. Which <u>table relationships</u> were changed (primary, foreign keys, natural keys)?
5. What data <u>source-to-target mappings</u> were changed?

Regression testing is important for data load processes whether accomplished through Informatica, stored procedures, or another process.

To plan for ETL regression tests, testers should identify how tables relate to each other (i.e., through examination of a data model) and use the knowledge, along with user specifications, to determine exactly what data warehouse data should be identical or changed across test ETL runs, leaving out change-prone values such as surrogate keys. The method provides tools for quickly detecting and displaying differences between the new ETL results and the reference results from an earlier date. In summary, manual work for test setup is reduced to a minimum, while still ensuring an efficient testing procedure.

What should be regression tested for ETL software is the result of the entire ETL run, or, in other words, the obtained side effects, not just individual function values. But even when the data is fixed in the input sources for the ETL, some things may change.

For example, the order of fetched data rows may vary in the relational model. Additionally, attributes obtained from *sequences* may have different values in different runs. However, actual values for surrogate keys assigned values from sequences are not interesting, whereas it indeed is interesting how rows are *connected* with respect to primary key/foreign key pairs.

For instance, it is not interesting if an ID attribute *A* has been assigned the value 1 or the value 25 from a sequence. What is important is that any other attribute that is supposed to reference *A* has the correct value. Further, the results to compare from an ETL run have a highly complex structure as data in several tables has to be compared.

Optionally, the user may specify joins, tables, and columns to include/ignore in the comparison. A QA diff process can then generate the so-called *reference results*, an offline copy of the DWH content. A query or series of queries can be run against tables before the change has been implemented. I take a snapshot of a table (ex. Day, week), then export to txt file for input to Excel or MS Access.

Whenever the ETL software has been changed, reference results can be compared with the current results, called the *test results*, and any differences will be pointed out. In the use case described above, the tuning team can use Excel or MS Access as labor-saving regression testing tools. Queries can be easily set up in MS Access to compare earlier data with new data to determine if results are as expected.

A tester will typically exercise the following tasks:

1. Write a SQL expression that joins the relevant tables and selects the columns to compare.
2. Verify that the query is correct and includes everything needed in comparisons.
3. Execute the query and write the result to a file.
4. Write an application or export to Excel/Access to compare results and point out differences.

Further, the user can exercise the following tasks for each new ETL version:

5. Run the new ETL software.
6. Run the query from step 1 again.
7. Start the application from step 4 above to compare the results from step 6 to the file from step 3. Even though much of this could be automated, it would take even more work to set this up. Thus, to set up regression testing manually takes hours, maybe days instead of minutes.

Comparing Data, Before and After Changes

When comparing data, there should be two data sets to consider. The results of an ETL run (the *reference results*) and the current result (the *test results*), that is, before and after. Methods to determine differences perform two tasks when running a test:

1) Export data from the DWH, and
2) Compare the test results to the reference results and point out any differences found.

A simple but effective and efficient strategy to retest basic functionality is to store source data sets and results from successful runs of the code and compare new test results with previous runs. When doing a regression test, it is much quicker to compare results to a previous execution than to do an entire data validation again.

Thoughts on Automating Data Warehouse Testing

A characteristic of data warehouse development is the frequent release of increasingly high *quality* data for user feedback and acceptance. At the end of each iteration of data warehouse ETLs, data tables are expected to be of sufficient quality for the next ETL phase or final production deployment. This objective requires a different approach to quality assurance methods. Foremost, it means integrating QA efforts and automation into our ETL iterations.

Essential to integrated ETL testing is test automation. Manual testing is not practical in a highly iterative and adaptive development environment. There are two key problems with manual testing.

First, it takes too long and is an inhibitor to the delivery of frequent working software. Teams that rely on manual testing ultimately end up deferring testing until dedicated testing periods, which allows bugs to accumulate.

Second, it is not sufficiently repeatable for regression testing. While we seek to embrace and adapt to change, we must always be confident that features that were "Done, done!" in a previous iteration retain their high quality in light of the changing system around them. Test automation requires some initial effort and ongoing diligence, but once technical teams get the hang of it, they can't live without it.

Integrated automated testing in database development presents a unique set of challenges. Current automated testing tools designed for software development are not easily adaptable to database development, and large data volumes can make automated testing a daunting task. Data warehouse architectures further complicate these challenges since they involve multiple databases (staging, presentation, and sometimes even prestaging); special code for data extraction, transformation, and loading (ETL); data cleansing code, and reporting engines and applications.

Data warehouse test automation can be described as the use of tools to control (1) the execution of tests, (2) the comparison of actual outcomes to predicted outcomes, (3) the setting up of test preconditions, and (4) other test control and test reporting functions. Commonly, test automation involves automating a manual process already in place that uses a formalized testing process.

Although manual ETL tests may find many data defects, it is a laborious and time consuming process. In addition, it may not be effective in finding certain classes of defects. Data warehouse test automation is the process of writing programs to do testing that would otherwise need to be done manually. Once tests have been automated, they can be run quickly and repeatedly. This is often the most cost effective method for a data warehouse that may have a long maintenance life because even minor patches or enhancements over the lifetime of the warehouse can cause features to break which were working at an earlier point in time.

Test automation tools can be expensive; they are usually employed in combination with manual testing. It can be made cost-effective in the longer term, especially when used repeatedly in regression testing.

What to automate, when to automate, or even whether one really needs automation are crucial decisions which the testing (or development) team must make. Selecting the correct features of the product for automation largely determines the success of the automation. Automating unstable features or features that are undergoing changes should be avoided.

Today, there is no known commercial tool not set of methods/processes that can be said to represent total end-to-end data warehouse testing. Informatica, Microsoft SSIS, and other major ETL tools do not promote any one tool as the answer to data warehouse automated testing.

Common Data Warehouse Test Automation Objectives

- Automate as many data warehouse testing cycle activities as possible.
- Verify data across all points during the ETL process.
- Verify all data, not just a subset or sample.
- Verify complex data transformation rules.
- Reduce manual testing workload which could be thousands of SQL scripts.
- Identify mismatched and missing data.
- Profile the performance of your data warehouse.
- Focus on numeric financial information for compliance or financial reporting.
- Reconcile and balance information between input and transformation processes.
- Validate information between the source and its final point in the data warehouse.

- Provide assurances to auditors that all financial information residing within a data warehouse can be trusted.
- Develop a regression test suite for system testing and production monitoring.

Automation of these checks may also be required due to the complexity of processing and the number of sources that require checks. Without automation, these checks would be costly or in some cases impossible to do manually.

Data Warehouse Processes—Targets for Automation

- File/data loading verification
- Data cleansing and archiving
- Load and scalability testing
- Application migration checks
- Extract, transform, load (ETL) validation and verification testing
- Staging, ODS data validations
- Source data testing and profiling
- Security testing
- Data aggregation processing
- BI report testing
- End-to-end testing
- Incremental load testing
- Fact data loads
- Dimension data loads
- Performance testing
- Regression testing

For most data warehouse development, the ETL test processes are primarily responsible for the data quality. Hence the selection of an appropriate ETL tool (e.g., Informatica, in-house developed) is a serious concern for an organization.

ETL tools and processes may contain a variety of the following components or systems. Most are reasonable targets for automated testing.

- *Aggregate building system*: For creating and maintaining physical database structures.
- *Backup system*: Responsible for backing up data and metadata.
- *Cleansing system*: Commonly a dictionary driven system for information parsing. For example, names and addresses of individuals and organizations, etc.
- *Data change identification system*: Used to record source log file reads, source date, and sequence number, etc.
- *Error tracker and handling system*: For identifying and reporting ETL error events.

- *Fact table loading system*: It is equipped with push/pull routines for updating transaction fact tables.
- *Job scheduling system*: For scheduling and launching all ETL jobs.
- *Late arriving fact and dimension handling system*: For insertion of fact and dimension records that because of some reason have been delayed in arriving at the data warehouse.
- *Metadata manager*: For assembling, capturing and maintaining all ETL metadata and transformation logic.
- *Pipelining system*: Required for implementing and streaming data flows.
- *Quality Checking System*: Responsible to check the quality of incoming data flows.
- *Recovery and restart system*: Responsible for restarting a job that has halted.
- *Security system*: Responsible for the security of data within an ETL.
- *Source extract system*: Includes source data adapters along with push/pull routines for filtering and sorting source data.
- *Surrogate key pipelining system*: Typically a pipelined, multithreaded process for replacing natural keys of incoming data with data warehouse surrogate keys.

Examples of Scenarios Used by Automated ETL Tools

Automated data warehouse testing tools may operate with the following sequence of tasks.

1. Client routine sends request to server to execute test.
2. Server routine assigns an agent to retrieve results from the source system for the test.
3. Server assigns another agent to retrieve the corresponding results from the target system.
4. The initial agent notifies server that the source information is retrieved.
5. The other agent notifies the server that the destination result is retrieved (may complete before number above).
6. Server assigns any available agent to compare the results and report any errors.
7. Agent returns to the server the outcome of the test case.
8. Server notifies all clients of the results of the test.

Deciding Which ETL Components to Automate

The trick is to figure out what needs to be automated, and how to approach this task. An array of questions must be considered in the course of automating tests such as:

- How much will it cost to automate tests?
- Who will do the automation?

- What tools should be used?
- How will test results be displayed?
- Who will interpret the results?
- What will script maintenance consist of?
- What will happen to the test scripts after the launch of the application?

Not all the areas of DWH testing can be automated, but some of critical testing involving the data synchronization can be achieved using automating data warehouse solution.

A primary objective of automated data warehouse testing is to cover the most critical part of data warehouse which is synchronization or reconciliation of source and target data.

Automation Challenges

- Large number of tables and records.
- Multiple source systems involved.
- Testing the data synchronization for all these tables.
- Testing Business Objects reports through automation.

Common Solutions

A start at ETL test automation can be divided into three major areas.

1. *Count Verification*: Initially all the tests for source to target mapping and Incremental Load for verifying initial/incremental/differential data count can be automated.
2. *Data Verification*: Secondly, for source to target mapping and Incremental load data verification tests can be automated.
3. *Verifying Data Types*: Lastly, logic for checking of the data types of source and target fields can be introduced.

Advantages

Reusability: These automated tests can be reused with minimum efforts for regression testing when newer versions of the source systems are introduced.

Automated testing tools are most helpful when used in the following circumstances:

- *Repetitive, mundane tasks.* As ETL processes are developed, they may go through many versions, each of which must be tested. By creating test scripts once, automated testing allows reuse for future versions.
- *Regression testing.* Each time a change is made to data or business rules, testers using a manual approach have to go through each function and make sure it does not affect any other part of the system. Automated regression testing uses test cases to do this quicker.
- *Multiple combinations* In increasingly complex environments where many scenarios must be examined, automated testing can rapidly run many test case combinations using a single script.

The Benefits of Automated ETL Testing

- *Executes test cases faster.* Automation can speed up test case implementation
- *Creates reusable test suites.* Once test scripts have been created in an automated tool, they can be saved for easier recall and reuse, and multiple testers often will have the capability and skills to run them.
- *Eases report generation and records test results.* An attractive feature of many automated tools are their ability to generate reports and test records. This capability can provide an accurate representation of the state of the data, clearly identify any defects, and be used in compliance audits.
- *Reduces personnel and rework costs.* Time that would be spent on manual testing or retesting after fixing defects can be spent on other initiatives within the IT department.

Limitations to Automated Testing

- *Does not eliminate manual testing.* Although automation can be used for many test cases, it can not totally replace manual testing. There are still some complex cases where automation may not catch everything, and for user acceptance testing, end users must manually run the test. Therefore, it is important to have the right mix of automated and manual testing in the process.
- *Cost of tools.* Automated testing tools can be costly, depending on their size and functionality. At a first glance, the business may not see this as a necessary cost; however, the reusability alone can quickly turn it into an asset.
- *Cost of training.* Some automated tools can be complex to use and may require training to get started. Testers must be trained not only on the software, but also on the automated test planning process.
- *Requires planning, preparation, and dedicated resources.* The success for automated testing is highly dependent on clear requirements and careful test case development before testing starts. Because each organization, scenario, and application can be unique, an automated testing tool will not create test cases. Unfortunately, test case development is still a manual process.

Recommendations

Evaluate the current state of testing in your organization. Not all ETL testing require automation. Assess the situations mentioned above to determine what type of automation would benefit the testing process and how much is needed. Evaluate testing requirements and identify inefficiencies that may be fixed with automated testing. QA teams that spend a lot of time on regression testing will benefit the most.

Make the business case for automated testing. Automated testing generally comes at a high price tag, so in order to convey the value to the business, the IT must first make the case.

Evaluate the options. After evaluating the current state and requirements within the IT department, look into which tools fit the organization's testing processes and environments. Options may include vendor, open source, in-house, or a combination of the aforementioned tools.

Bottom Line

Automated ETL testing tools can significantly reduce the amount of time spent testing code in comparison to traditional manual methods. Other benefits include the creation of reusable code and a reduction in costs associated with personnel and rework. Get educated on automated testing and available tools to decide if it's right for the QA team.

A Sampling of Automated Testing Scenarios

A tenet of agile and other modern development is automated testing. We can apply this idea to our data warehouse as well.

An important consideration about testing the data mart is that the number of tests that are run during the testing phase continues to grow. As time passes, we should have more tests testing more functionality.

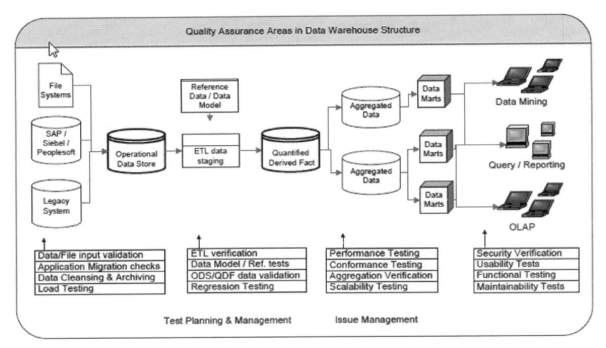

Points of ETL Testing

As shown above, there are four major points of ETL testing. For data automation testing, the emphasis at each test entry point is the validation of data integrity. Using test automation, data can be tracked from the input layer, through the ETL processing and into the XML messages produced (in this case), onto storage in the data warehouse and finally to the front-end applications or reports. If corrupt data is found in a front-end application or report, the execution of automated suites can help rapidly determine whether the problem is located in the data source, an ETL process, in a data warehouse database, or in the front-end itself.

For performance testing, the same test entry points are utilized to focus on characterizing subsystem response under load. A similar strategy is used for data integrity validation to determine the origin of performance issues. That is, subsystem performance can be measured at any of the identified test entry points. This is especially significant with regard to the test entry point located at the data warehouse component itself. This entry point tests one of the key architectural features of a data warehouse—its optimization of queries.

Data warehouses commonly use strategies such as aggregate tables, partitioning and B-tree indexing in order to speed up queries, and the success of the architecture used should be directly verified. The emphasis on rapid localization of either data or performance problems in complex data warehouse architectures provides a key tool for promoting efficiencies in development, and for shortening build cycles and meeting release targets.

Predeployment ETL layer testing can be summarized as follows:

- Input synthesized feeds (flat files, feeds, other XML grammars) of known data content.
- Validate output data (which may be stored in XML, database or other data formats).
- Validate data integrity at post-ETL entry points, e.g., if the processed data is persisted in a database, validate the database contents based on input data.
- Validate all data conditions for all feeds.
- ETL layer performance assessment.
- ETL subsystem/component performance assessment.

Predeployment data warehouse and front-end testing can be summarized as follows:

- Validate user queries based on a data warehouse of known data composition.
- Validate user queries for all front ends in use.
- Performance of data warehouse.
- Performance of the front-end applications.

Some Vendor's Test Automation Solutions

RTTS currently markets QuerySurge; InfoSys markets Clearware. These are among the few test tools offering some level of end-to-end ETL test automation. With these tools, users will find that among the tasks is to rebuild the programmer's (or ETL tool's) data extraction SQL, which is a manual process. If the tester's data sets don't match the DWH, there may be a problem. As users manually build the SQL, it is plugged it into an automated test that actually compares every record and column data value on millions of rows, and tells us where the differences are. The byproduct is an automated regression test.

QuerySurge™ is built around a central repository that stores queries, allowing testing of millions of rows of data during a run. Once the QuerySurge repository is populated with queries that are customized for each source system and each business rule, volume data testing can begin. The tool framework is fairly simple—it pulls queries from the repository and executes them against the source systems and the data warehouse.

The queries are written to return the data in a common format, regardless of differences in database structure. The tool then compares the results of these queries to find discrepancies in the data within the different systems.

Infosys Clearware, a data warehouse testing solution, is a comprehensive tool designed to test high-volume data. It automates data warehouse test processes, enables extensive reuse of test components, ensures early detection of data defects, and provides complete traceability. Clearware follows a well-structured model to validate data, enhance quality, and accelerate the testing process.

Sure, rebuilding much of what the programmers did is challenging. Automation testers will need to learn case statements, cross table joins, data type casting, and keep their heads straight while working through queries that include multiple more joins of various kinds.

Highlights of a Comprehensive DWH Test Strategy

Testing Goals and Verification Methods

Primary goals for DWH verification over all testing phases include:

- *Data completeness* Ensure that all expected data is loaded.
- *Data transformation* Ensure that all data is transformed correctly according to business rules and/or design specifications.
- *Data quality* Ensure that the ETL application correctly rejects, substitutes default values, corrects or ignores and reports invalid data.
- *Performance and scalability* Ensure that data loads and queries perform within expected time frames and that the technical architecture is scalable.
- *Integration testing* Ensure that the ETL process functions well with other upstream and downstream processes.
- *User-acceptance testing* Ensure the solution meets users' current expectations and anticipates their future expectations.
- *Regression testing* Ensure existing functionality remains intact each time a new release of code is completed.

Data Completeness

One of the most basic will be to verify that all expected data loads into the data warehouse. This includes validating that all records, all fields and the full contents of each field are loaded. Strategies include:

- Comparing record counts between source data, data loaded to the warehouse and rejected records.
- Comparing unique values of key fields between source data and data loaded to the warehouse. This is a valuable technique that points out a variety of possible data errors without doing a full validation on all fields.
- Utilizing data profiling methods that show the range and value distributions of fields in a data set. This can be used during testing and in production

to compare source and target data sets and point out any data anomalies from source systems that may be missed even when the data movement is correct.

- Populating the full contents of each field to validate that no truncation occurs at any step in the process. For example, if the source data field is a string (30) make sure to test it with 30 characters.
- Testing the boundaries of each field to find any database limitations. For example, for a decimal (3) field, include values of -99 and 999, and for date fields, include the entire range of dates expected. Depending on the type of database and how it is indexed, it is possible that the range of values the database accepts is too small.

Data Transformation

Validating that data is transformed correctly based on business rules is a complex part of testing ETLs with significant transformation logic. One method is to pick sample records and *stare and compare* to validate data transformations manually. This can be useful, but requires manual testing steps and analysts who understand the ETL logic. Following are some of the simple data movement techniques:

- Create a spreadsheet of scenarios of input data and expected results and validate these with business SMEs. This is a good requirements elicitation exercise during design and can also be used during testing.
- Create or identify test data that includes all scenarios. Elicit the help of ETL developers to automate the process of populating data sets with the scenario spreadsheet to allow for flexibility because scenarios will change.
- Utilize data profiling results to compare range and distribution of values in each field between source and target data.
- Validate correct processing of ETL-generated fields such as surrogate keys.
- Validate that data types in the warehouse are as specified in the design and/ or the data model.
- Set up data scenarios that test referential integrity between tables. For example, what happens when the data contains foreign key values not in the parent table?
- Validate parent-to-child relationships in the data. Set up data scenarios that test how orphaned child records are handled.

Data Quality

Data quality here is defined as "how the ETL system handles data rejection, substitution, correction and notification without modifying data." To ensure success in testing data quality, include as many data scenarios as possible. Data quality rules will be defined during design, for example:

- Reject the record if a certain decimal field has nonnumeric data.
- Substitute null if a certain decimal field has nonnumeric data.
- Validate and correct the state field if necessary based on the zip code.
- Compare product code to values in a lookup table, and if there is no match load anyway but report to users.

Depending on the data quality rules of the application being tested, scenarios to test might include null key values, duplicate records in source data and invalid data types in fields (e.g., alphabetic characters in a decimal field). Review the detailed test scenarios with business users and technical designers to ensure that all are on the same page. Data quality rules applied to the data will usually be invisible to the users once the application is in production; users will only see what's loaded to the database. For this reason, it is important to ensure that what is done with invalid data is reported to the users. These data quality reports present valuable data that sometimes reveals systematic issues with source data. In some cases, it may be beneficial to populate the *before* data in the database for users to view.

Performance and Scalability

As the volume of data in the data warehouse grows, ETL load times can be expected to increase, and performance of queries can be expected to degrade. This can be mitigated by having a solid technical architecture and good ETL design.

The aim of the performance testing is to point out any potential weaknesses in the ETL design, such as reading a file multiple times or creating unnecessary intermediate files. The following strategies will help discover performance issues:

- Load the database with peak expected production volumes to ensure that this volume of data can be loaded by the ETL process within the agreed-upon window.
- Compare these ETL loading times to loads performed with a smaller amount of data to anticipate scalability issues. Compare the ETL processing times component by component to point out any areas of weakness.
- Monitor the timing of the reject process and consider how large volumes of rejected data will be handled.
- Perform simple and multiple join queries to validate query performance on large database volumes. Work with business users to develop sample queries and acceptable performance criteria for each query.

Integration Testing

Usually system testing only includes testing within the ETL application. The endpoints for system testing are the input and output of the ETL code being tested. Integration testing shows how the application fits into the overall flow of all upstream and downstream applications.

When creating integration test scenarios, we will consider how the overall process can break and focus on touch points between applications rather than within one application. Consider how process failures at each step would be handled, and how data would be recovered or deleted if necessary.

Most issues found during data integration testing are either data-related to or resulting from false assumptions about the design of another application. Therefore, it is important to integration test with production-like data. Real production data is ideal, but depending on the contents of the data, there could be privacy or security concerns that require certain fields to be randomized before using it in a test environment.

To help bridge this communication gap between IT and the business, we will gather team members from all systems together to formulate and verify test scenarios and discuss what could go wrong in production. Run the overall process from end to end in the same order and with the same dependencies as in production. Integration testing should be a combined effort and not the responsibility solely of the team testing the ETL application.

User-Acceptance Testing

The main reason for building our data warehouse application is to make data available to business users. Users know the data best, and their participation in the testing effort is a key component to the success of a data warehouse implementation. User-acceptance testing (UAT) typically focuses on data loaded to the data warehouse and any views that have been created on top of the tables, not the mechanics of how the ETL application works. Consider the following strategies:

- Use data that is either from production or as near to production data as possible. Users typically find issues once they see the *real* data, sometimes leading to design changes.
- Test database views comparing view contents to what is expected. It is important that users sign off and clearly understand how the views are created.
- Plan for the system test team to support users during UAT. The users will likely have questions about how the data is populated and need to understand details of how the ETL works.
- Consider how the users would require the data loaded during UAT and negotiate how often the data will be refreshed.

Regression Testing

Regression testing is revalidation of existing functionality with each new release of code and data. When building test cases, they will likely be executed multiple times as new releases are created due to defect fixes, enhancements or upstream systems changes. Building automation during system testing will make the process of regression testing much smoother. Test cases should be prioritized by risk in order to help determine which need to be rerun for each new release.

A simple but effective and efficient strategy to retest basic functionality is to store source data sets and results from successful runs of the code and compare new test results with previous runs. When doing a regression test, it is much quicker to compare results to a previous execution than to do an entire data validation again.

Purpose of the Test Strategy

The purpose of the test plan is to ensure that the project, which includes a data conversion, is thoroughly tested, resulting in a successful implementation of new and existing functionality, business processes, reports, interfaces, and batch processes.

The functional area test plan consists of the following:

A. Test Roles and Responsibilities
B. Items To Be Tested
C. Test Strategy
D. Test Approach
E. Test Readiness Assumptions Met
F. Deliverables
G. Approval

Table 1: Test Roles and Resources

Role	Responsibilities—related to testing
Test Designers: 1. Devin Smith, Data Quality, ETL Tester	• Develop and document test design for data conversion, functional, life cycle, security, and performance testing. • Create and document test cases based on test design. • Coordinate testing activities within own functional area. • Communicate test updates to Test Coordinator and Functional Project Managers.

Tester: 1. Devin Smith, Data Quality, ETL Tester	• Run test cases during designated test periods. • Document test results and problems in quality center. • Work with developers to troubleshoot problems. • Retest problem fixes. • Communicate test updates to test leads.
Project Manager: 1. Project Mgr	• Monitor and update project plan testing activities by functional area. • Facilitate in the development of test plan, test design, and test cases for functional and life cycle testing.
ETL Developer: 1. Par 2. Dan	• Perform unit test. • Troubleshoot problems. • Document problem updates in test director. • Communicate unit testing progress to test lead. • Work with test designers/testers on data validation planning and testing efforts.
DWH Test Coordinator Devin Smith, Data Quality and ETL Tester	• Develop master test plan. • Prepare test materials (test guidelines, procedures, templates) for functional areas use. • Setup test and defect reporting. • Setup Test Director for testing effort. • Monitor test planning progress and execution efforts. • Facilitate and coordinate the development and execution of cross functional area testing. • Redevelop automated tests for post implementation. • Work with Functional Project Managers on test plan tasks. • Keep functional areas and project team leads appraised of testing activities and test results.

Items To Be tested

Tests will be developed for the list of items below. Details of the items to be tested should be provided in the test design and scenario document—a future deliverable.

Tests from the Functional Area

1. Data Validations

   • Transformation values
   • Set-up tables
   • Row counts for each table

- Detailed review of specific cases (Transactions, Products, Adjustments to be determined)

These steps can be exercised for each source-to-target data load

- Verify all the source tables have been imported and loaded to targets.
- Verify all the rows in each source table have been imported and loaded to target.
- Verify all the columns specified in source table have been imported and loaded to target correctly. Sample data to assure no dropped or altered values. Verify that sums of individual columns match source.
- Verify all the data has been received without any truncation for each target column.
- Verify the data schema at source and destination to include data formats, data lengths, data type, and precision is as defined in data model.
- Verify the time taken /speed for data transfer is acceptable.
- Verify that range of values for each field (min, max) meet expectations.
- Verify no duplicate rows in target.
- Verify no duplicate data in columns which should be distinct.
- Verify that each transformation was done according to specifications for each row in the target table.
- Verify no failure in referential integrity, foreign key to primary key.
- Verify that all integrity constraints on source data were applied as expected.
- Review ETL error log for rejected records and other errors.
- Verify precision of data for each numeric field.
- Verify that not null fields are not null or blank.
- Verify that primary keys were developed and there are no duplicates.
- Verify that triggers, procedures, and functions are migrated successfully. Check that the correct values are returned for triggers and functions.
- Verify that default values have been applied to fields where defined and under defined circumstances.
- Verify that calculations or manipulations associated with data are performing as expected and yield accurate results.

	Test Description	Steps	Input
1	Verify the account loaded	See Table 1	See source-to-target mapping document
2	Verify the adjustment reason loaded	See Table 1	See source-to-target mapping document
3	Verify the account loaded	See Table 1	See source-to-target mapping document

	Test Description	Steps	Input
26	Verify successful completion of each ETL package		
27	Identify ETL packages that ended with errors		
28	Verify that expected SSIS or SQL Server error messages are issued upon specific error conditions		
29	Verify each defined ETL exception processing routine.	See "ETL Exception Processing" worksheet.	
30	Testing ETL Error Exception Processing		
31	During extraction—verify handling of wrong or unexpected data in the table.	For example, place the wrong SSN format, character fields in what should be numeric	
32	Verify ETL handling: Not Null - Source column is null while target is not null		
33	Verify ETL handling: Reference Key - The records coming from the source data do not have a corresponding parent key in the parent table.		
34	Verify handling: Unique Key - The record already exists in the target table.		
35	Verify handling: Check Constraint—Check constraints enforce domain integrity by limiting field values.		
36	Nonfunctional Data Testing		

	Test Description	Steps	Input
37	Performance Testing (i.e., tests that ensure the performance, such as average response times, remains within acceptable tolerance levels),		
38	Connectivity Testing (i.e., tests to ensure the effectiveness of connectivity components, such as networks and servers, is unchanged by the data changes),		
39	Operations Testing (i.e., tests of components, such as job control language),		
40	Storage Testing (i.e., tests to determine the requirements for production size data stores),		
41	Stress Testing (i.e., tests to ensure that the execution of peak volumes, representative of production)		
42	Interface Testing (i.e., tests that evaluate the accuracy of interfaces with temporary data <u>bridges</u> and after temporary bridges are removed).		

Table 2: General Test Scenarios

2. Business Processes (Including Business Requirements and Business Rules)

- Search for a Rep online (by ID, name, CRD, SSN)
- Add person online
- Attempt to create Rep = 'NEW' online
- Update person online—name, addresses, phones, e-mail, biographic data
- Change national ID online
- Try to add duplicate national ID online
- Validate a person's data affiliations online
- Personal Portfolio tests (if this component is implemented)
- Manage national IDs for multiple countries (if this is implemented)

3. Reporting

4. ETL Packages

- SEVIS alerts
- Linked addresses update
- Campus housing purge
- Post OIS holds for new admits (if process is still needed)
- Purge OIS batch holds (if process is still needed)

5. Interfaces

6. Cross Functional Areas Processes (Life cycle Testing)

7. Functional Area Key Process Performance (Timing Measurements)

- Online person search
- Online person add

Load Testing

No load testing is currently planned.

Security Testing

- Change RepCode (function is tied to a specific security class)
- Demographic Data Access (DDA) testing
- Personal Portfolio security (if this component is implemented)

Test Planning Readiness Assumptions and Needs

A first level of testing and validation begins with the formal acceptance of the logical data model and low-level design (LLD) documents, including ETL design and functionality implemented in SSIS. All further testing and validation will be based on the understanding of each of the data elements in the data model and ETL design.

For example, data elements and related fields that are created or modified through transformations or summary process must be clearly identified and calculations for each of these data elements must be unambiguous and easily interpreted.

During the LLD reviews and updates, special consideration should be given to typical modeling scenarios that exist in the project. Examples follow:

1. Many-to-many attribute relationships are clarified and resolved.
2. Types of keys used are identified: surrogate keys versus natural keys.

3. Business analysts/DBA reviewed with ETL architect and developers (application) the lineage and business rules for extracting, transforming, and loading the data warehouse.
4. All transformation rules, summarization rules, aggregation, matching, and consolidation rules are documented.
5. ETL procedures are documented to monitor and control data extraction, transformation and loading. The procedures should describe how to handle exceptions and program failures. The ETL design document describes implementation of all the above.
6. Data consolidation of duplicate or merged data is properly handled.
7. Target data types, precision, field lengths are as specified in the design and/or the data model.
8. Mandatory (not null) indications are available for all fields.
9. Fields are initiated with, default, null or blank values where expected.
10. Default values are specified for fields where needed.
11. Acceptable values are provided for each field.
12. Expected ranges for field values are specified where known.
13. Slowly changing dimensions are described.

Test Strategy

Listed below are the various types of testing that will take place in four test phases.

Testing	Role	Description
Data Validation	ETL Developer, Tester	Validating the data converted to PS 8 database. This is done early in the project.
Unit	ETL Developers	Developer testing to validate that new code can be executed to completion with no errors.
Functional	Tester	To validate the application meets business requirements under various scenarios for a functional area.
Performance	Tester	Measuring the time an activity or key process takes to complete. This testing should be done throughout the test phases.
Load	Tester	Running key processes or activities under heavy usage simulation to determine if the application can perform adequately with high levels of activity/transactions.
Regression	Tester	Regression testing to validate PS 8 patch and updates.
Acceptance	Tester	Customer testing of the application for approval for production. This testing typically is not in-depth testing.

Testing	Role	Description
Security	Key Security Contact	Testing security implemented.
Data Warehouse (DWH)	IS DWH Tester	Testing to validate changes to DWH. This would involve IS DWH testing the DWH data stage jobs in the conversion and the testers validating the repository queries.
Automated	Test Coordinator	Run redeveloped automated tests

Test Approach

The four test phases are based on the development schedule for PS 8 upgrade project along with the requirement to comply with financial aid regulation updates that need to be in place when the upgrade goes live.

Phase 1:

- Data Validation
- Performance
- Unit
- Functional
- Data Warehouse (internal testing within IS validating data stage jobs)

Data validation should start early in the test process and be completed before phase 2 testing begins. Some data validation testing should occur in the remaining test phases, but to a much lesser extent.

Important business processes where performance is important should be identified and tested (when available) in the phase 1. Performance testing should be continued in the later test phases as the application will be continuously enhanced throughout the project.

In addition to phase 1 testing, there will also be unit and functional testing. As unit testing is completed for a program, the tester will perform functional tests on the program. While functional testing takes place with one program, the developer continues with redeveloping and unit testing the next program.

Toward the end of phase 1, the data warehouse group will be testing the data stage jobs. Redevelopment and unit testing should be completed then functional testing finishing a couple weeks afterward. A final formal test will cap the end of phase 1 testing.

<u>Phase 2</u>:

- Cross-functional process
- Load
- Security
- Data warehouse (Repository testing and validation)

In addition to the above tests, phase 2 should also cover remaining test items that may not been tested in phase 1 such as:

- Business processes
- Cycle

Phase 2 testing will be important because it is the final testing opportunity that is and the functional area testers will have to make sure the DWH load works as expected before moving to regression testing in phase 3. Some performance tests and data validation should be included in this phase.

A final formal test will cap the end of phase 2 testing.

<u>Phase 3</u>:

- Regression

Phase 3 testing is comprised of three regression test periods to test update patches and regs that are required as part of the Go-Live system in late 2012. The functional area testers will have two weeks to test in each regression test period.

<u>Phase 4</u>:

- Business and client Acceptance

Phase 4 testing is limited to one or a few weeks. In addition to the functional area testers, end users will probably be involved in this final test before the system goes live.

In customer acceptance testing, no new tests should be introduced at this time. Customer acceptance tests should have already been tested in prior test phases.

Data Track QA Entry Criteria

1. ETL unit tests are complete to include DB load verification in DEV.
2. ETL code, DB issues, changes requests, and open defects from development have been documented.

3. Data track release notes have been delivered and no critical ETL or DB issues are open.
4. QA has access to test/DEV environment for minimum of one week before QA start date.
5. Approved data model and mapping document (LLDs) available minimum of two weeks before QA begin (required to prepare test cases).
6. ETL and DB QA test cases complete and in QC.
7. All QA environment schema are set up according to the test plan for current build/iteration.
8. ETL code migrated from development environment in QA environment.
9. QA was provided all access to DB and ETL code in QA region of one week before QA start date.
10. Developers finished ETL sanity check successfully in QA region of three days before QA start date.

Schedule

Test Activities	Role	Start Date	End Date
Create Master Test Plan	Test Coordinator		
Create Functional Test Plans	Test Coordinator		
Create Test Design	Test Designer		
Create Test Cases	Test Designer		
Phase 1 Testing	Testers		
Phase 2 Testing	Testers		
Phase 3 Testing	Testers		
Redevelop Automated Test Cases	Test Coordinator		
Create Customer Acceptance Criteria	Test Coordinator		
Phase 4 - Customer Acceptance Testing	Testers		

Deliverables

A. Functional Test Plan
B. Functional Test Design
C. Functional Test Cases

Risk Management for the Data Warehouse

The greatest risk in life is not taking any risks.

—Robin Sharma, *The Monk Who Sold His Ferrari*

This is a story that may have not happened, but ought to have happened, of how a great company was shattered into mediocre existence by its own success.

Once upon a time, not so long ago, there was a great company great to its customers, great to cities and countries wherever it operated its business.

People all around the world admired this company. It was more than a company; it was an institution known for its passion to be right, its rigorous processes, its obsession for outguessing customers' desires, its thorough training of employees, and its "cradle-to-grave" employment policy. Yet the very same attributes that made the great company admired worldwide soon became its shortcomings.

The proclaimed values that made the company famous could not sustain it a world where the marketplace was changing at the speed of internet. Soon the walls it had built to protect its employees became walls of imprisonment for those who remained within them for too long. The company became known for its aversion to risk-taking, its brainwashing of employees, its failure to invent new products and its inability or unwillingness to anticipate its customers' desires.

The company created a comfort zone for its professional employees that made them unwilling, unable or afraid to move out it. Occasionally, some employees would snap out of the realm of their comfort zone, but people above and below would pull them back into their midst until eventually they succumbed to the rules of its civil service mentality.

We interviewed some ex-employees of this virtual company with the simple question:

> Why did the great company collapse?

The answer was unanimous—the upper echelons were to blame:

> Senior management was fat and happy; the rank-and-file didn't have much say in the matter.

This statement, taken at its face value, is true. After all, noted authority Dr. Edward Deming stated convincingly that 96 percent of all the problems in a company can be attributed directly to its management. The more interesting question is, *Why* didn't the leaders of the great imaginary company do their job? Presumably everyone wants to do a good job. The upper echelon is not an exception to this rule. The only logical explanation is they did not know how!

Two developments back in 1960, more than anything else, shaped the future of the information technology industry as we know it today. First was the invention of microprocessors which led to the development of personal computers (PC) and the second was the development of graphical user interface (GUI).

Modern GUI was derived from PARC User Interface (PUI, also an acronym for *perceptual* user interface), developed by researchers at Xerox PARC (Palo Alto Research Center). PARC user interface consisted of graphical elements such as windows, menus, radio buttons, check boxes, and icons, and it also employed a pointing device in addition to a keyboard.

In 1979, Apple Computers, started by Steve Jobs with some former members of the Xerox PARC group, continued to develop the ideas of the perceptual user interface. The Macintosh, released in 1984, was the first commercially successful product to use a GUI desktop metaphor in which files resembled pieces of paper; directories resembled file folders, and accessories appeared as calculators, notepads, and alarm clocks. The user could manipulate those metaphors around the screen as desired and could, for example, delete files and folders by dragging them to a trash can on the screen.

In the meantime, our great imaginary company—even though it had been using its own microprocessors many years before they were discovered by the rest of the industry, and even had a great pool of talented software engineers, who developed prototypes using this new technology—stood by doing nothing. Its upper management lacked vision; they just could not imagine the future with a PC on every desk and its impact on business and society in general. They could not see emerging patterns for new business opportunities in which every employee would be empowered by the technology they already possessed in their labs and would be able to utilize their extensive knowledge and expertise of their employees. This coupled with their aversion for risk taking prevented them to seize upon an opportunity.

The processes of seeing patterns of the future, making new connections and risk taking to seize opportunities, are basically creative right brain activities, and the more you try to structure and organize them, the more they disappear. Our imaginary company was run by left brain leadership, which invested much energy into structure and organization, and owed its success to left brain thinking. As a result, it was unable to switch gears overnight, reinventing itself into a new paradigm of whole brain thinking that was appearing behind the horizons.

In the collision between the creativity of the right brain and the logic of the left brain, subservient right brain thinking loses based on the final argument: "This is not the how the things are done around here." The more dominant left brain thinking prevails, by stamping over someone's dreams that ware like a delicate spring flower so innocently planted in mud before their feet. Eventually, we learn the lesson of never planting flowers anywhere.

Continuing these causal analyses, the same question remains: "But *why*?" It turns out that the problem runs much deeper into our society. Picasso once said that all children are born artists. When we send them to school, some are labeled as having learning disabilities.

This is the story of a six-year-old girl who was labeled as being unable to concentrate. In her drawing class, she was sitting in the last row, deeply concentrated in her drawing. The teacher came to her desk and asked what she was drawing. Without taking her eyes of her work, she answers:

> "I am drawing a picture of God."
> "How do you know, what God looks like? Nobody has seen Him?"
> "They will in a minute!" she answered.

Later, her parents explained it to her that if she wants to qualify for university and ultimately get a job, she must concentrate on math, science and English. This is how society eventually educates our children out of their natural creativity. By avoiding experimentation and risk taking, especially out of fear of what others may think, it puts a limitation on our creative capability, as nothing new can be invented if one is not prepared to take a risk of making mistake. Our educational system is broken, and attempting to improve on something that is already broken is not going to correct the quintessence of the problem. We know the education system is broken otherwise we would not be continuously trying to improve it. We are even giving some fancy name to those programs, like "No Child Left Behind," but sadly, the more things change, the more they look the same. Radical new thinking is required, as Albert Einstein suggests:

"We can't solve problems by using the same kind of thinking we used when we created them."

The educational system narrowly focused on academic ability; inevitably marginalizes students with interests and gifts in other domains. As Socrates expressed it:

"Education is the kindling of a flame, not the filling of a vessel."

Learning is a very personal process. Each student has different needs and learning styles. The purpose of teaching should be to ignite the students' imagination by supporting her or his passion and learning style. No one can be forced to learn against his or her free will. Yet that is precisely what is being done currently. Students are compelled to learn outside their natural mental environment and ultimately pay the penalty of failure. For example, even the most reluctant students eventually learn to commit facts and ideas to memory, or drop out of the system.

Partly to blame for the adaptation of this faulty educational model is renowned Swiss developmental psychologist Jean Piaget, who maintained that intuitive ability was prevalent only during the first two years of life, but afterward is overtaken by other powerful abstract and intellectual ways of *knowing*. His stage theory of development had a profound impact on several generations of educators, who saw it as their job to redirect children's learning away from the reliance on their senses and their intuition, while, at the same time, encouraging them to learn skills of deliberation and explanation sooner rather than later.

There is ample evidence that this kind of thinking is widespread in society and even more so in the corporate world (with some exceptions). For example, a sophisticated mathematical model consisting of hundreds of partial differential equations of the national economy may be represented by a computer program which can measure everything that can be counted. Conversely, anything that has no measure, like human nature, which is intangible (impossible to measure), has no value. In other words, our assessments are like those of a man who was searching for his car keys under a streetlight, only because that is the only place where he could see.

Inherent in this mode of thinking is linguistic inadequacy of language used in describing business requirements, especially requirements for developing DWH application. The fundamental rule of writing business requirements is that if something is to be understood, it must be written clearly and unambiguously in order for the rational (logical) mind to understand and ultimately create a computer program. Language inadequacy poses a problem when it comes to describing evocative world of metaphors and imagery, most prominently exposed in describing DWH application requirements. These kinds of problems require different types of *knowing*; *knowing* that emerges from not knowing, uncertainty, and the mind's acceptance of this transitional state as those are the seeds from which creative thinking will sprout.

Western culture has lost the sense of this different ways of *knowing*. These different ways of *knowing* are referred to as subconscious intelligence and conscious intelligence. Conscious intelligence is simply referred to as *intelligence*, thus denying even the existence of any other intelligence. In making sense of the world, our common thinking process is dependent on pattern recognition based on our prior experiences. But to find something new, we cannot depend on quick decision (rational thinking) as they are inhibitors to deeper intelligence (subconscious intelligence) since they are based on previous experience.

Thinking creatively occurs when the mind slows and relaxes, thus activates a new and a unique pattern of neurons, forming novel association between them, only then the other ways of *knowing* automatically appears. William (Bill) Lear world famous American inventor and aeronautical designer, attributed for saying that he had no education and that was why he had to use his brain, also said:

> "One of the unfortunate things about our education system is that we do not teach students how to avail themselves of their subconscious capabilities."

The new discipline of *cognitive science* association in the fields of neuroscience, philosophy, artificial intelligence, and experimental psychology is confirming the existence of the subconscious intelligence of the human mind and that this subconscious intelligence is responsible for accomplishing the most unusual tasks—analyze and make sense of the most complex situation, providing it had enough time to do so.

The most ingenious solution does not come about as a result of logical reasoning, they "occur to us." Complex mantel processes occur without our control or awareness. To the famous Italian opera composer Giacomo Pucini, while composing his opera *Madame Butterfly*, it appeared that

> it was dictated to me by God; I was merely instrumental in putting it on paper and communication it to the public.

On the other hand, if education was personalized to match each student's needs, fewer students would pull out of it. Some say that personalized education is an impossible pipe dream, as it would too expensive. But we argue that it would be more expensive not to. The fundamental point of reference must be that education is not a cost, but investment in our future. Based on the second premise that all students are curious by nature, it is clear that and that most efficient and profound learning takes place when it is initiated and pursued by the learner. All the students are creative if they are allowed to develop their unique talents within the appropriate educational environment, and if we want to conceive future opinion leaders, there must be a better ways to educate our children.

There is another more pragmatic reason for unleashing the artistic right brain power within an organization. Artists, those heralds of the future, create not only for the present generation, but even for those that have not been born yet. A good artist is many years ahead of his or her time, or in the words of the great Canadian hockey player, Wayne Gretzky: "I am where pack is going to be, not where it has been."

The artists' way of knowing is not a special privilege of artists and sages; it is available and could be cultivated by the rest of us. Once when we were very young, we knew that, and it is not that we have forgotten it; it is just that we don't trust it any longer.

The new business reality requires all of us to awaken the artistic power within organizations, which up till now has been suppressed and unwelcome. Leading MBA programs are introducing art and design in their curricula. Leader of companies must learn new skills if they want to conceive the new products to be ahead of the game, rather than waiting for their competitors' products to recognize the future.

If the above conclusion seems farfetched, then let's take a look at some examples of successful products in the marketplace today. The iPod, for example, was most certainly developed employing a minimalistic approach to design of the user interface. This example is evidence that the art and patterns of expansive thinking are nurtured in an organization where the collective aspiration is set free, as an integral part of the corporate culture.

A similar example of this kind of the organization is Google. The word *Google* has become a new verb and not just in the English language. Companies that embrace art as the part of their corporate culture are prospering, while their closest competition is continuously playing a catch-up game.

In many situations, our rational intelligence is necessary for our survival, yet there are always other ways to perceive and act then through our habitual way of thinking. Creative thinking is increasing freedom in our lives by providing other options and opportunities in problems resolution. Creativity is not about discovering new lands, but looking at the same land with a new set of eyes. It is about seeing the extraordinary within the ordinary. As the Tao Te Ching puts it:

> "Truth waits for eyes unclouded by longing.
> Those who are bound by desire see only the outwards container."

Risk is the possibility of suffering loss. Business must learn to anticipate and mitigate the possible negative consequences of risk against the potential benefits of its associated opportunity. In itself, risk is not bad; in fact, it is essential to progress. Failure is often a key part of learning too. Risk taking and the significance of learning from failure are often overlooked. Corporate culture that nurtures creativity is also tolerant to risk taking as it sets the stage for innovative thinking. Ingredients inherent to this culture that are, among other things, necessary for creativity, are the ability to challenge assumptions, recognize patterns, seeing out of the box, make new connections, take risks and seize the moment upon a chance.

The future is not the place we will go to, aimlessly like a ship with no rudder on the open ocean. The future is a place the leaders of a company must invent and clearly communicate and steer an organization in that direction.

Putting the blame on the educational system is convenient scapegoat for abolishing personal responsibility, as schools should not be looked upon as the end, but the beginning of education. With or without the education, each one of us has to make a choice, whether we want to live today as yesterday waiting for Fridays, or do we engage in adventurous journey into unknown full of possibilities, which eventually becomes that of what we think about. Thus, proving that we are the masters of our destiny, as one thing that is always under our control, is what is going on in our heads.

QA Strategy Review Time

Person can have the greatest idea in the world—completely different and novel—bat if that person can't convince enough people, it doesn't matter.

—Gregory Berns[9]

A QA strategy document is an important document, but in this case, it differs from company's standard practices and procedures, and it may be in grave danger if it is not it not effectively announced to the world. The stakeholders must be informed and convinced before they can commit to full support and funding required for the project to proceed. If there is no support for it, it will go into the waste basket, where many documents that cross our decks end up.

The message of the QA Strategy is too important to be wasted. But the strategy document is hampered by one-dimensionality of written word. Socrates (Greek: Σωκράτης) one of the founders of Western philosophy, did not write anything. He believed that writing is inferior form of communication in comparison to the power of oral argument. Written documents could not answer questions therefore could not stand up for themselves. The opportunity to present QA Strategy document is QA Strategy Review session, which is a required event in many organizations. Full advantage of this event must be taken, by effectively presenting the QA Strategy story. You have the podium and it's your show. Use it—but wait, not before you read the next section.

---

[9]   Gregory S. Berns is a distinguished American neuroscientist professor of psychiatry and writer.

The Power of Storytelling

People will forget what you said, people will forget what you did, but people will never forget how you made them feel.

—Maya Angelou[10]

Imagine human beings living in a underground cave, which has a mouth open towards the light and reaching all along the cave; here they have been from their childhood, and have their legs and necks chained so that they cannot move, and can only see before them, being prevented by the chains from turning round their heads. Above and behind them a fire is blazing at a distance, and between the fire and the prisoners there is a raised way; and you will see, if you look, a low wall built along the way, like the screen which marionette players have in front of them, over which they show the puppets. The wall conceals the puppeteers while they manipulate their puppets above it. Imagine further man behind the wall carrying all sorts of objects along its length, and holding them above it. The objects include human and animal images made of stone and wood and all other material.

Imagine now one of the prisoners is freed from shackles. He turns around and walks towards the lights. He suffers pain and distress from the glare of the lights. So dazzled is he that he cannot even discern the very objects whose shadows he used to be able to see. He is blinded by light, actually by the truth, but unable to comprehend it.

He is dragged away by some force up the steep incline of the cave passageway, until he hauled out into the light of the sun.

He goes back to the cave, where the others are shadow watching to reveal the truth to them. The brave sole who risked all to seek the truth is now being ridiculed for even attempting the journey!

This, of course, is the famous *Allegory of the Cave*, extracted form Plato's *The Republic*. The *Allegory of the Cave* represents an extended metaphor for the state of human existence, and for the transformation that occurs during enlightenment. By using storytelling, Plato hopes to enable the reader to embrace the full potential that has been living beneath the surface. *When the light of the sun shines on the freed man* is the metaphor for enlightenment and perception of the truth.

---

[10]   Marguerite Ann Johnson is an author and poet who has been called "America's most visible"

In the *Allegory of the Cave*, Plato brings to light the idea that without education, in this case through stories, people are like prisoners in a cave living amidst shadows. People without knowledge know nothing about experience or desire. Their reality consists of the images that dance before them on the walls of their cave. Yet this cave is not a physical one; this cave is an imprisonment of the mind. To Plato, the most important thing in life is to be educated continuously in the hopes of better understanding life and all that is a part of it.

Upon returning to the cave, our escaped prisoner would metaphorically be entering a world of darkness again, and would be faced with the other chained prisoners. The other prisoners would laugh at the released prisoner, and ridicule him for even taking the useless ascent out of the cave in the first place. The others cannot understand something they have not experienced, so it's up to this prisoner to epitomize leadership, for it is him alone who is conscious of goodness. It's at this point that Plato describes the philosopher kings who have seen the Ultimate Goodness as having a sense of duty to be responsible leaders and to not feel disdain for those whom don't share his enlightenment.

Most people, including ourselves, live in a world of relative ignorance. We are even comfortable with that ignorance, because it is the best we know. When we first start facing truth, the process may be frightening, and many people run back to their old ways. But if you continue to seek truth, you will eventually be able to handle it better. In fact, you want more of it. It's true that many people around you now may think you are weird or even a danger to society, but you don't care. Once you've tasted the truth, you won't ever want to go back to being ignorant.

The aspect of the above story we are interested in here is: Now that you have seen the *light*, how do you move other out of their *comfort zone* and how do you confront all those *I-told-you-so-ers* and *naysayers*, who will offer many reasons why that won't work, and who will kill you at the end if you don't convince otherwise.

The prescription for the solution to this dilemma came from Plato's student, Aristotle. Plato devised and used his *Allegory of the Cave* as a teaching tool, and Aristotle, who ought to have been part of those discussion, turned to the most ancient art of storytelling, to resolve the dilemma, of how to return to the cave and convince the fellow cave dwellers that at the end of that long passage into their cave, there is the most beautiful world—a promised land.

In his book *The Poetics*, Aristotle has revealed the *secret* of story-telling to us, more than twenty three centuries ago. The story has to appeal to emotions (*pathos*), reason (or logical appeal—*logos*), and personal credibility (ethical appeal—*ethos*). Facts alone are not sufficient to persuade. The most effective way to convey the most complex idea is to turn it to a story.

Sufi mystics have for centuries instructed their followers with stories. Many Sufi tales have been passed down the centuries into legends and ethical teachings. The tale of the elephants used earlier in *Story of Quality Assurance* section of this book is a famous example. The Islamic tradition of Sufi tales entertains while teaches at the same time. The goal of stories is to amplify perception and helping to gain deeper knowledge. Only a few Sufi stories can be read by anyone and still unlock internal consciousness. Sufi masters unlock the internal dimension of the stories when he feels their disciples are ready.

Stories have been used since time immemorial for communicating important messages, to inspire or share knowledge and ideas. Everyone responds to a good story. For as long as humans have existed, it has been the way of relating, history, legends, and lessons—sometimes to entertain, or even to save lives as is an old Persian story of Scheherazade (Farsi: داستانهای هزارو یک شب شهرزاد قصه گو). This ought to be a true story as she lived to tell it to us:

Scheherazade was a Persian queen and the narrator of One *Thousand and One Nights* (also called *The Arabian Nights*). The legends of the Arabian Nights were passed down through the centuries by word of mouth; the oldest tales date to the tenth century.

It is about a Sultan who, upon discovery of his wife in orgy with her slaves, vows to never trust women again. To protect himself from the future pain, he was in the habit of marring a virgin and killing his wives after the first night. This was going on for three years, until Scheherazade conceived a plan to stop him and convinced her father to offer her as the sultan's next wife.

Scheherazade was the oldest daughter of the grand vizier to Sultan Shahryar. Scheherazade was very well-educated young lady, having studied the legends, books, histories, and stories about kings and civilization in general. She had learned all about philosophy, poetry and arts. Not only was she well read, but also was well-bred, polite, and pleasant to all she encountered.

The night after their wedding ceremony, Scheherazade told Sultan her first story. Sultan Shahryar listened in awe as Scheherazade spun a fascinating, adventurous tale, but she stopped speaking before the story was finished. The Sultan asked her to finish the tale, but Scheherazade said there was no time left because it was almost dawn and time for her beheading. She added that she really regretted not finishing this story, because her next story was even more thrilling.

Sultan Shahryar decided not to execute Scheherazade that morning just so he could hear the rest of the story later that night. Scheherazade began one of the exciting tales but stopped before the story ended, causing the sultan, who had listened as well, to put off killing her until she could finish her story the next evening. Scheherazade, of course, never finished her tales, but kept her husband enthralled with story after story for 1,001 nights.

Scheherazade must have known that here storytelling was not to show up how brilliant she was, but about making Sultan the hero. While presenting her stories, she was humble, almost invisible, as her success depended on the Sultan, not the other way around.

At the end of one thousand and one nights and one thousand stories, she knew she had reached remarkable outcome. Scheherazade told the Sultan that she had no more tales to tell him. During these one thousand and one nights, the Sultan had fallen in love with Scheherazade, and had three sons with her. Having been made a wiser and kinder man by Scheherazade and her tales, he not only spared her life, but made her his queen.

Storytelling in Business

This report, by its very length, defends itself against the risk of being read.

—Winston Churchill

Despite of what we have been told, the greatest invention of all time isn't the wheel; it's business organization. Communication is life blood of a business organization. Effective communication is essential for the prosperity and survival of a business. It is also a basic tool for motivation and education of the employees in an organization. The power of storytelling is such a potent communication tool, that it makes the need to learn how to do it better, a must for business professionals. Every great story is about problems and problem resolution because that is very much what the life is all about. The story of QA Strategy is not different; it presents the problem and its resolution.

In business, we are exposed to a variety of stories on a daily basis, yet we do not fully appreciate the power of *storytelling* in reframing people's viewpoints. How many times at business meetings have one or more participants insisted that each had the correct perception of the issue at hand? Arguments are prolonged as they try to defend their respective positions, unwilling to entertain the possibility of an alternative point of view or to see the problems through a different lens. There is also the resistance to change, the resistance to seek new methods to resolve business problems because participants are often caught up in the message of the old stories which have worked so well in the past. It is very much like the diagram of the cube below. Once we see it from one perspective, it is difficult to see it from another perception.

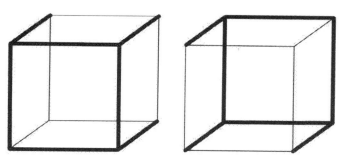

Fig. 8.1 Cubes

PowerPoint presentations have become the norm in business today. During a typical PowerPoint presentation, the speaker stands before an audience and reads out the presentation bullet point after bullet point. This style of presentation is the main reason why many business presentations are dismal failures; they also out of sync with how people communicate and learn.

Edward Tufte, the visual communication guru and author of wonderful books like *Beautiful Evidence* and *Visual Explanation,* puts the blame directly on PowerPoint. In the September 2003 issue of *Wired Magazine*, in the article titled "PowerPoint Is Evil," Tufte wrote,

> A presentation format should do no harm. Yet PowerPoint style routinely disrupts, dominates, and trivializes the content. Thus PowerPoint presentations too often resemble a school play—very loud, very slow, and very simple.

Dr. John Sweller of the University of South Wales, who developed the cognitive load theory[11], discovered that it is more difficult to process information if it is simultaneously both verbal and visual. In other words, people cannot listen and read at the same time. The assumption made is that the PowerPoint presentation, which includes lines of text projected on a screen, mirrors the spoken words. As a result, reading PowerPoint slides to an audience can be ineffective in communicating a message. Poorly designed presentations do not achieve the presenter's intention and are not an enjoyable or effective learning experience for the audience.

Making a presentation should not be viewed as an opportunity to for the presenter to demonstrate his *brilliance*. Rather it is an opportunity to make a small contribution or to share an idea that that the presenter feels is important. Millions of mediocre PowerPoint presentations are delivered every day, but the good news is that the bar

---

[11] The term **cognitive load** is used in cognitive psychology to illustrate how the load is related to the executive control of working memory. People learn better when they can build on what they already understand. The more a person has to learn in a shorter period, the more difficult it is to process that information in working memory

is now so low that there is plenty of opportunity to make a difference with even a small improvement. A presentation that goes well will have a strong impact on the presenter's spirit, confidence, and career. A good presentation means the *story* had been told effectively.

> *The single most important thing you can do dramatically improve your presentation is to have a story to tell before you work on your PowerPoint file.*

> —Cliff Atkinson[12]

Despite the overwhelming evidence that presentations using PowerPoint (or Keynote) should be abandoned, we will prove, by method of exhibiting an example, that PowerPoint presentations can still be effective.

Steve Jobs's 2007 iPhone launch[13] presentation is an excellent example. It is a clear, visually supported story in which today's world and his vision of an improved future world are contrasted. His presentation consists of slides without bullet-points. Slides contain only pictures to support Jobs's story.

It is the *story* and the metaphors used in the story that convey the message, bullet-points on a PowerPoint (Keynote, in this case) do not engage conceptual thinking. Why is this? The audience responds best to tangible, moveable, and inspirational things. Humans make decisions based on emotion, and then look for the facts to support their decisions or choices. Facts alone just can't do that.

This is supported with the recommendations of the latest findings in cognitive theory. The research of Dr. Richard Mayer, educational psychologist at the University of California, which has been reported in a study named *A Cognitive Theory of Multimedia Learning* in which he outlines the fundamental principles of multimedia design based on cognitive functioning. He writes,

> It is better to present an explanation in words and pictures than solely in words … When giving multimedia presentations, present corresponding words and pictures contiguously rather than separately … When giving multimedia explanations, use fewer rather than many extraneous words and pictures."

2-PAGE SHORT STORY in TWO DAYS?

---

[12]   Cliff Atkinson, *Beyond Bullitt Points*
[13]   Steve Jobs announcing the new iPhone. Apple Reinvents The Phone

Mark Twain, the prolific American writer, received this telegram from his publisher:

NEED 2-PAGE SHORT STORY TWO DAYS?

Twain replied:

NO CAN DO 2 PAGES TWO DAYS. CAN DO 30 PAGES 2 DAYS. NEED 30 DAYS TO DO 2 PAGES.

A shorter presentation with relevant information is more in tune with cognitive learning theories. Interestingly, this is also a fundamental Zen concept known as *kanso* or simplicity. This reminds us of the Japanese minimalistic approach discussed earlier in this book. Japanese Zen arts claim that it is possible to exhibit great beauty and convey powerful messages through *kanso*. (Jobs's message of "a thousand songs in a pocket" sounds very much like a *kanso* message).

This is the same principle supported by Dr. Edward Tufte in eliminating *chartjunk*— eliminating superfluous design elements. On the same topic, German painter Hans Hofmann said:

The ability to simplify means to eliminate the unnecessary so that the necessary may speak.

The *minimalistic aesthetic* approach is evident in all Apple products. Time and time again, we see the same approach in Steve Jobs's presentations.

One of the most important parts of Apple's design process is simplification.

—Leander Kahney[14]

Indeed, the art of storytelling is an art form which, more than any other art form is integral to the human experience. When presentations are delivered in a story framework, they put on view in context of the subject of the delivery, making it a formidable motivator for the audience.

However, few of us have the public-speaking confidence of Steve Jobs. Knowing the rules of the art does not necessarily make an artist a *presenter*. It should not come as a surprise that confidence is the result of hours of relentless practice. In fact, confidence is so important to the quality and effectiveness of a presentation that US president Barack Obama once said that the most important lesson he learned from

---

[14]   Leander Kahney, *Inside Steve's Brain*

working with the most powerful person on the planet was to "always act confident." Ultimately, the result of countless hours of rehearsal in any field of endeavor is confidence and effortless performance.

Storytelling is an art that can be learned. Even the world's geniuses, such as Wolfgang Amadeus Mozart, who composed from the age of five and performed before European royalty, had to invest hundreds of hours to master his art form. In his book *Genius Explained*, psychologist Michael Howe[15] explains that to be a genius demands a strong sense of direction and an extraordinary degree of commitment, focus, practice, grueling training and drive.

> "By standards of a mature composer, Mozart's early works are not outstanding. The earliest pieces were all probably written down by his father, and perhaps improved in the process. Many of Wolfgang's childhood compositions, such as the first seven of his concertos for piano and orchestra, are largely arrangements of works by other composers. Of those concertos that only contain music original to Mozart, the earliest is now regarded as a masterwork (No-9, K.271) was not composed until he was twenty-one: by that time Mozart had already been composing concertos for ten years."

If Mozart had not persisted as long as he did, the world would have never known the beautiful arias from "The Marriage of Figaro" or "The Magic Flute." We continue to enjoy the benefits of his work today; according to a landmark neuroscience research study out of University of California[16], listening to one of the most profound and most mature Mozart's composition "Sonata for two pianos Concerto, K. 488" produces significant short-term enhancement of spatial-temporal reasoning in college students.

Persistence and perseverance require passion and continuous encouragement for achieving mastery in one's chosen calling. For that reason, it is important for each of us to discover our true passion as it is an absolute prerequisite for stretching our imagination and reclaiming our brain creative power and putting in service to whatever we value most and, in that process, realize our full potential.

First, we need to search for what is truly beautiful, whether it is a perfect piece of poetry, or piece of music, or laughing with a loved one and experiencing the deepest feelings of empathy. We need to associate with good people who tell good stories about other good people because the people we surround ourselves with have a

---

15   Michael J.A. Howe is professor of psychology at the University of Exeter
16   Center for the Neurobiology of Learning and Memory, University of California, Irvine, Department of Physics, University of California, Irvine, CA 92717, USA 18 October 1994

profound effect on who we are and who we become. In other words, we need to learn to relate to other human beings in a positive way.

A beautifully told story is a symphonic unity in which structure, setting, character, genre, and ideas melt seamlessly. In order to find their harmony, the author must study the elements of the story, as if they were instruments of an orchestra—first separately, then in concert. Life is too short to be wasted in associating with people who dismiss our passion and discourage our ideas.

Second, the relentless practice for many hours over many days is required to pursue and eventually to master our life's calling. Neuroscientist Daniel Levitin claims that the magic number in achieving mastery in any task is ten thousand hours:

> "The emerging picture of such studies is that ten thousand hours of practice is required to achieve the level of mastery associated with being a world-class expert in anything . . . In study after study of composers, basketball players, fiction writers, pianists, chess players, master criminals, and what have you, this number comes up again and again. Of course this does not explain why some people don't seem to get anywhere with their practice and why some people seem to get more out of their practice session than others, but no one has yet found a case in which true world-class expertise was accomplished in less time. It seems that it takes the brain this long to assimilate all that it needs to know to achieve true mastery."[17]

To put this period 10,000 hours in prospective, it is equivalent to 10 years of performing the task repeatedly three hours per day. We now have the scientific proof for what we have known for years, and indeed as old adage goes: practice make perfect (Latin *Repetitio est mater studiorum*). Rehearse and rerehearse! This is not different with skill of communication, as the presentation can not be better than the preparation effort that went into it.

In his book *Brain Rules*, John Medina[18] (under rule number 5—"Repeat to remember"[19]) claims that it is that period of approximately ten years of repeating a task that is required for transferring temporary memories into more persistent forms.

Medina told the story of an experiment in which a patient's hippocampus was removed because of a life threatening disease. After the operation, the patient could not recollect anything from the last ten years of his life, but he had a clear memory

---

[17] Daniel Levitin: *This is Your Brain on Music*

[18] John Medina: Brain Rules: 12 Principles for Surviving and Thriving at Work, Home, and School

[19] From the book *"Brain Rules"*—Rule #5: Repeat to remember.

of events from eleven years and earlier prior to the removal of the hippocampus. This proves that the patient's temporary memory, not yet committed to his more persistent forms of memory, was completely erased by removing patient's hippocampus.

Story-telling language is metaphor as that is the script language that directly communicates with the conceptual mind. The essence of metaphor is to understand one unfamiliar concept by relating it to the familiar one. Learning of new concept is enabled by quickly associating it to a familiar knowledge. The new idea is experienced by relating it to a familiar one. It's the conceptual system that defines the reality. All communication is interpreted by the same conceptual system which is also used in thinking and acting. Therefore if we want our audience to quickly grasp the contents of our story, we must use metaphors.

The presenter's objective is to fully engage brains of the audience (stakeholders). Before making a presentation, the presenter should know something about the audience, their thinking styles, extraverts, introverts, detail-oriented, conceptual types, etc. More importantly is to know what makes them tick in order to connect with them.

The presenter's goal is to unify the group of individual who have temporary come together to hear presenter's story. In order to make the audience focus to the presentation, the presenter must learn something about them, their skills, a few names, their weaknesses, demographics, etc. as this is where the empathy for the audience will come from. Otherwise, as Ken Haemer of ATT says:

> "Designing a presentation without an audience in mind is like writing a love letter and addressing it: To Whom It May Concern."

Based on the above discussion, an effective presentation must still follow the story line prescribed by Aristotle many centuries ago—a well-designed story must have a beginning, a middle, and an ending. The beginning sets the stage for the story where the emotional contact is made with the audience and where the problem is introduced. The middle part of the story develops the action where the main protagonist (hero) is faced with obstacles which prevent him or her from achieving a desired outcome. The ending is reserved for the resolution of the problem.

Every presentation requires a blueprint—a destination on a road map to show how to reach the desired outcome, the place where the audience's understanding and commitment are intended to be following the presentation. If the audience does not understand the presenter's blueprint, they will soon disconnect from the conclusion of presentation, both mentally and emotionally. The presenter's goal is to transport the audience from one location to another, or to persuade them to let go of the old ways and adapt new ones. Two questions must be answered for the audience if the presenter wants win their attention: "What is your point?" and "What's in it for us?"

The beginning is the critical part of the story. This is where Aristotle's classic power of establishment is engaged appealing to emotions and reason based on personal integrity. This is where the protagonists and their relationships are described and the *hero quest* is established. The intent of the first slide of the presentation is to immediately draw in the audience and focus their attention on the presentation. The questions to answer at this point are "Where are we now?" and "Who are we in this setting?"

After making an emotional connection and presenting the problem clearly in the first part of the presentation, the audience is ready to hear the appeal to reason in the middle part of the presentation. They are at this point closely paying attention to learn how *they* can solve the problem. The how and the why provide justification for their decision to accept the proposed solution in the first part of the presentation and, at the same time, this is the reason why they have come to here the presentation.

Ending part of the presentation is where everything emotional and rational is tied together. The stage is set up for the audience—the protagonists of the story—to hear the answer to the question they have at this point: "What needs to happen so that we can resolve the situation?"

Journey through Bermuda Triangle

Visual analogy with the ships and the voyage through the Bermuda Triangle is the subject of the conceptual metaphor in the visual story.

To set up the scene, the title of this slide is: "The ships are safe in the harbor, but that is not what the ships are built for."

Visual: The body of the slide of a sailing ship docked in a tranquil harbor.

Narrative[20]: All organizations go through a life cycle of starting up, growing, maturing, and eventually declining unless, of course, they *reinvent* themselves. (This is known as a concept of S-curve: new businesses start out slow, and then scale rapidly and taper off). Successful businesses are founded on the idea of the world and the future world as an improved place. It requires courage and communication to recognize the declining cycle long before it is obvious to others. It also takes courage to move

---

[20] The narrative text is entered in the notes part of the PowerPoint slide. If you must read your notes, Apple's presentation software, Keynote, allows speakers to see the notes, while the audience sees just the slides displayed on the projector screen. But it is more important maintaining the eye contact with the audience to reading the notes. Research has shown that eye contact is associated with integrity and trustworthiness. Conversely avoiding eye contact is associated with lack of confidence and sincerity.

forward to an unknown future which can yield both risks and rewards. Regardless of the uncertainty of the future, businesses must move forward in order to survive. Companies must learn to live and prosper in the chronic tension of what is and what could be. We are in the middle of the paradigm shift.

This is not dissimilar to the stories of Bill Gates and Steve Jobs. Both of them toiled away in relatively obscure field of microcomputer without any hopes for any worldly success. But then it happened—the personal computer paradigm shift! They were ready; they had their 10,000 hours in. The world changed, and they led the change, while the major computer manufactures stood by, wandering what happened.

The second slide should introduce the *hero*. At this point, the audience will ask: "Who are we in this scene?"

The presenter is not the hero; the presenter is there to support the audience and to help them recognize that they are the *real heroes* of the story. This slide also introduces the adversity and as author Robert McKee, in his book *Story*[21], writes,

> "Something must be at stake that convinces the audience that a great deal will be lost if the hero does not obtain his goal."

So here it is, the title of the second slide which identifies the hero and the adversity is: "Future of your company is in your hands."

Visual: will have diagram of S-curve upon the S-curve, showing the paradigm shits.

Narrative: We today again have a choice, to maintaining the status quo by pretending that nothing is happening (and disappear into oblivion), or embrace the new BI reality. Inaction is not a smart choice, as we will soon find ourselves on steep declining curve, wandering what happened to our organization and consequently to every one of us.

After the paradigm shift, trying to climb the old S-curve will no longer work. We must make the jump from our current S-curve to a future S-curve now. If our intention is to compete in the future marketplace and prosper in it, perusing BI has no alternative. The quest for BI is long and never ending journey that we all have to make commitment to if we intend to find a new prosperity in the new marketplace on the personal level too!

The next slide title introduces the imbalance, with the title: "BI today is sailing through the Bermuda Triangle."

---

[21]   Robert McKee *Story: Substance, Structure, Style and The Principles of Screenwriting*

Visual: Ship zigzagging through the stormy sees toward its destination.

Narrative: But journey to prosperity leads trough some dangerous waters. In fact so dangerous that nine out of ten ships are lost forever. The problem is caused by proliferation of one-size-fit-all solution.

The cause of the problem is introduced in the next slide which should answer the question: "What stands between us and our dreams?" This also is the title in the heading in the next slide: "What stands between us and our dreams?"

Visual: Picture of summer sky in the evening with billions upon billions of stars.

The presenter's narrative begins: In summer evenings, humans have always looked up and wondered about the mystery of the billions upon billions of the stars in the sky about which we know so little. There are also that many billions of neurons flying around the most sophisticated machine in the universe, inside our brains. Yet we know very little about our brains. However, one thing we do know for certain is that intelligence is not *out there*; it is inside our brains just waiting to be unleashed.

Business or any other intelligence is simply not *out there*; rather, it has to occur inside that most sophisticated machine—our brains. And that is exactly what we want to do; by unleashing human intelligence, we achieve BI. There is nobody else from outside who understands our business better than we do, and who can do it better then we can do. The proliferation of the one size-fits-all solution is the cause of the problem and that by the way is the reason why business intelligence today is contradiction in terms, an oxymoron.

This will follow the slide stating the desired outcome: "In pursuit of wisdom—journey without ending."

The body of the slide (visual) reflects a zigzag path of the ship from the start point to the destination, and the additional path which the ship would have taken without onboard navigational system. This metaphorically relates the value of the prototype to an onboard navigational system. In other words, an onboard navigational system is enabling the crew to avoid the obstacles along the way, but at the same time, know where we are all along the path in relation to the desired destination.

Narrative: The secret is that the intelligence and wisdom to create BI for your organization is inside every one of you. Using previously discussed Excel prototype, based on something familiar, we will be able build something new—a BI for your organization. The awareness of knowing our location is continuously obtained from the business user, thus forming the coherent team of business and IT experts moving and learning together through an uncharted sea. In the process of moving and learning together, human intelligence is unleashed, a process that can never be even initiated by any third party software.

The learning process will be enhanced and eventually mastered by many repetitions of this cycle as we repeatedly navigate through the stormy seas of the business intelligence, similar to the repeated rehearsals of the masters in the cycle of 10,000 hours. The ending of one cycle is simultaneously the beginning of the next cycle of the journey for the BI team. And so it is repeated *ad infinitum*.

> What we call the beginning is often the end.
> And to make an end is to make a beginning.
> The end is where we started from

T. S. Eliot

It is much later than you think!

During the research for his book *Inside Steve's Brain*, Leander Kahney was struck by Jobs's apparent preoccupation with death, signified by how many times he mentioned it. This may be the driving force behind Jobs's urgency to "dent the universe." In a commencement speech to the graduating class at Stanford in 2005, he observed:

> "Remembering that you are going to die is the best way I know to avoid the trap of thinking you have something to lose. You are already naked. There is no reason not to follow your heart"

In the book you are reading, it's not by accident that the terms used throughout it, such as *battlefield, warriors, strategy, trenches, echelon*, etc., are all metaphors with one common denominator: the conceptual subject of those metaphors is war. Those metaphors *are* used in this book because they convey the urgency and importance of the subject and its consequences.

It's just that the process is too gradual and hard to be noticed, but the evidence is convincing: we are losing ten percent per year of well paid, *white collar* jobs every year; our houses are being repossessed at an alarming rate; soon we will not be able to afford our educational and health systems that we have benefited so far.

We are facing a *silent crisis*, in which things are changing in real and meaningful ways, but such changes are hard to detect in our everyday lives. Maybe you've heard of the experiment with a frog before, but let's use it here to make this point clearer:

Throw a frog in a pot of water. If you turn the heat up on the water quickly, the frog will sense the temperature change and jump out of the pot. But if you raise the temperature slowly, the frog won't notice until it's cooked, but then it's too late to do anything about it.

Pretending nothing is happening is not an astute strategy. First, we must find out what is the cause (not the consequences, as those are the just the symptoms) of the alleged problems described in the above paragraph. The author Thomas Friedman seems to know the answer.

In his best-selling book *The World Is Flat*[22], he attributes this to a phenomenon which he calls *globalization*. Friedman, while on a journey to India, realized that *globalization* has changed core economic landscape of the world. In his opinion, this *flattening of the world* is a result combination of number of coincidental events like crumbling of borders (collapse of the Berlin Wall), convergence of the advances in fiber-optic networking coupled with personal computer, etc.

Friedman's *Flat World* is probably his metaphor which relates to the work, which is becoming commoditized and sold in global markets. He should (but he does not) further refer to the commoditization of only the mundane work in manufacturing and services, or reading manuals at call centers over the phone, etc., but excluding not-so-mundane intellectual activities, like product innovations and high-paying, noncontestable creative jobs like those required in development of Business Intelligence.

It is neither globalization nor a *flat world* that is causing problems in our labor market (but it's a nice scapegoat). It is the paradigm shift. Developed countries are moving away from the "economics of goods" to the "economics of information." Its technology and the new requirements of post-industrial society that is causing labor market shift.

While it is true that many skills required for prospering in the industrial age are no longer required, this paradigm shift has created tremendous opportunities for those that were ready for it. Adaptation of deep new knowledge is required and the need to learn fast in order to stay relevant and prosper in the new labor market. It appears that paradigm shift rules apply equally well to everyone of as they do to organizations.

Instead of blaming our misfortune to circumstances beyond our control, it's better to embarrass the new reality as the old one is gone forever. The proof of this premise is that many jobs go unfilled due to lack of qualified professionals, this seem to

---

[22] *The World Is Flat: A Brief History of the Twenty-First Century* is an international bestselling book by Thomas Friedman

coincide with the opinion of qualified and respectable source in their MCKINSEY QUARTERLY MONTHLY NEWSLETTER JULY 2011[23]:

Big data: The next frontier for competition

The ability to work with vast data sets—big data—could spur productivity and innovation, but the United States alone has a shortage of up to 190,000 people with suitable training in statistics and machine learning. Our interactive exhibit, on the McKinsey & Company Web site, explores the industries and occupational categories where these specialists work.

On a final note, the reader should remember that these guidelines are only points to ponder and not mandates. Art can open established doctrines available not only to artists but engineers and programmers and it should be used by them when the circumstances dictate. Effective PowerPoint presentations exist as a new type of art, form heralded by interested presenters, and certain rules can and should be broken if the situation warrants a change.

"Just one last thing" (Jobs), remember the 10,000 hours rule; the sooner we start packing in those hours, the sooner we will become experts in BI.

---

[23] http://www.mckinsey.com/en/Features/Big_Data.aspx

Test Automation Case Study

By three methods we may learn wisdom: first, by reflection, which is noblest; second, by imitation, which is easiest; and third by experience; this is the bitterest.

—Confucius

Background:

A new store is opening at a new location. In order to manage the logistic and marketing information, business team wants to receive all the daily transactions and associated customer information from the store. The new ETL feed job will be created to transfer all the transaction data from the new store to the enterprise DWH. The ETL job will be running nightly at midnight. Files including data file and header-trailer information will be loaded from data warehouse landing area and copy to stage database. A nightly ETL job will load those data from stage database insert into master data warehouse and report database.

Test Case 1

Scope and Description of Data Warehouse of Testing: Verification of Source Data loading into Staging Database.

Objective: To verify data integrity of the transaction files based on the header and trailer information.

Testing assumption: The test script will be reused [company is going to open or relocate other stores], therefore, it is a good candidate for regression and test automation.

Related Data Source (shown on the picture below fig. 12.1)

Two file are created and placed on the FTP Landing server. Store Business Transaction Data and Header and Trailer files

(1) Incoming Data file [File name: IT_STORETRANSACTIONS.DAT]
(2) Incoming Header and Trailer file for Transaction Data [File name: TT_ STORETRANSACTIONS.HT]
(3) Incoming file Transaction Data file is loaded into Stage Database [Oracle 10g database]. Header and Trailer file is used for verification of Transaction Data file.

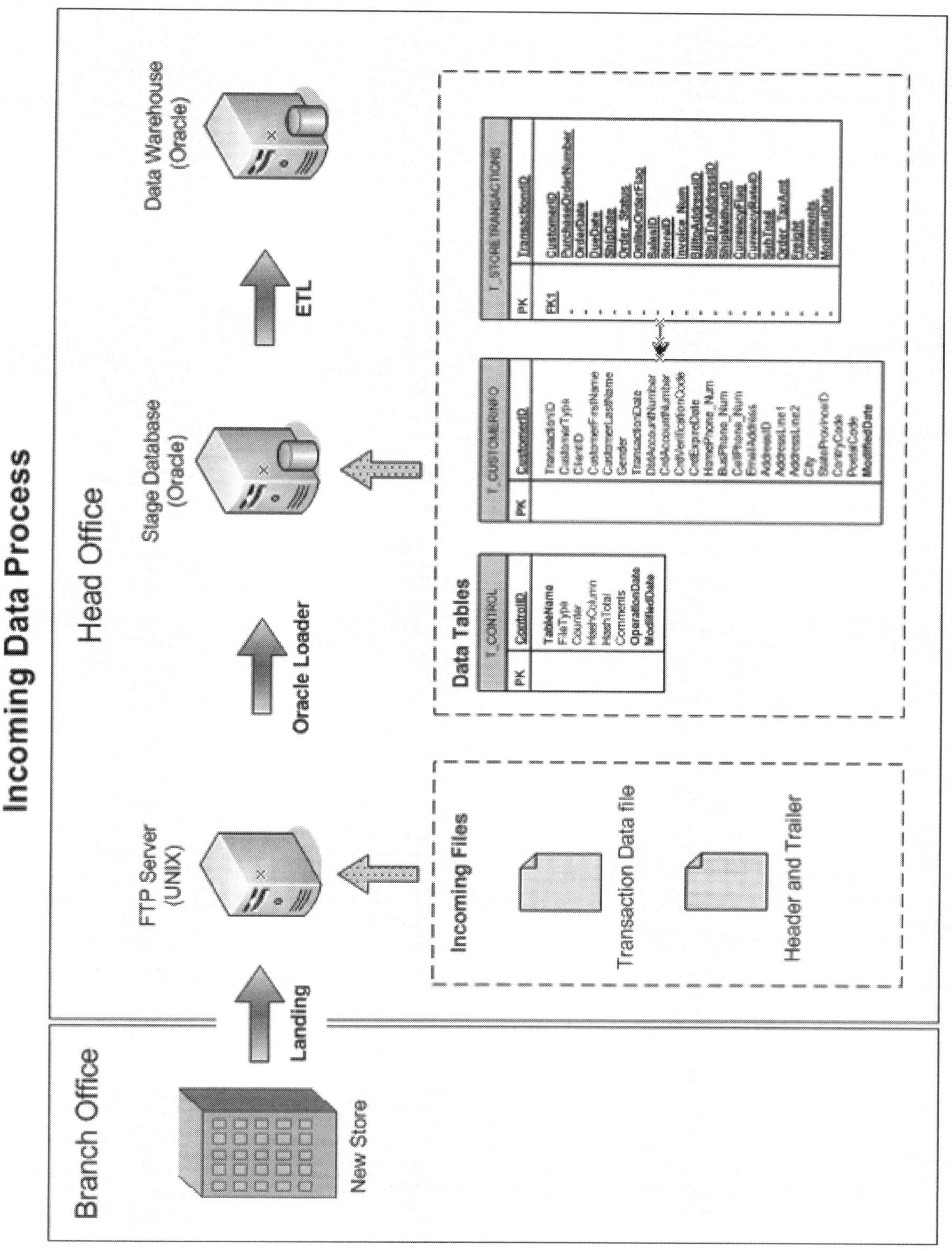

Incoming Data Process

Figure 10.1: The data incoming process

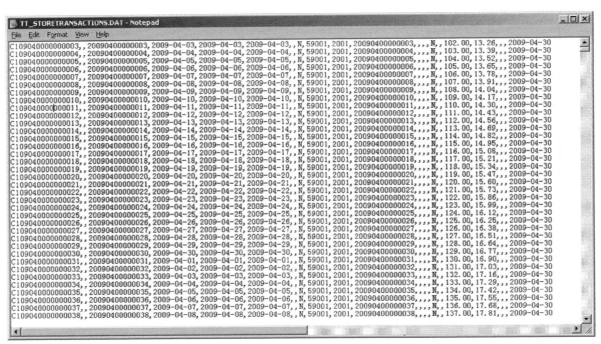

*Figure 10.2: The data file sample for incoming data source
—Store Transaction Feed.*

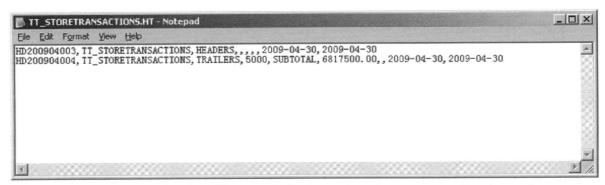

*Figure 10.3: The data file sample for incoming data source—Header and
Trailer.*

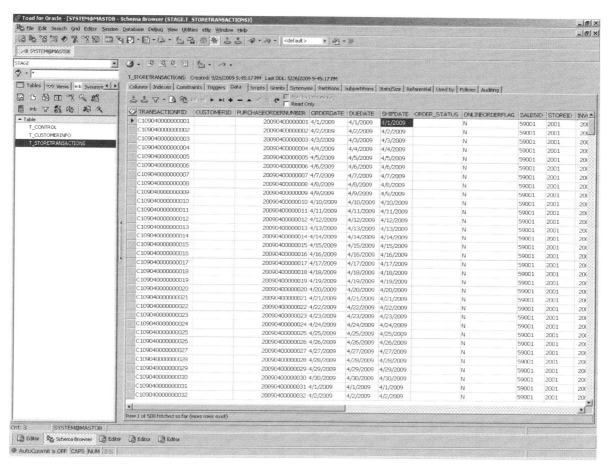

Figure 12.4: Incoming data populated in the Stage Database
(STAGE.T_STORETRANSACTIONS table).

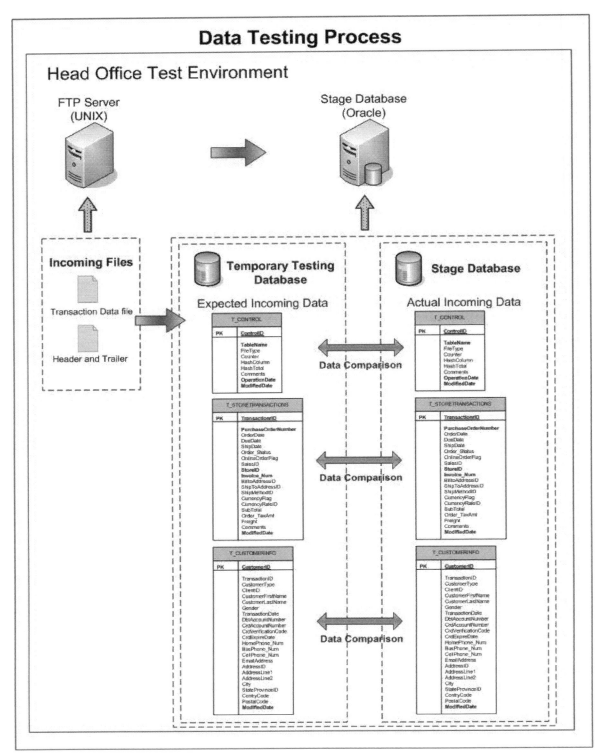

Figure 10.5: Transaction Data Verification Process

<u>Description of the verification process</u>:

Oracle Loader is used to load incoming store Transaction Data file into "TestUser" schema. Test table is created and Store Transaction Data is copied into the test table. In the last step of setting testing environment "test _result" table is created, to store the verification results.

In final verification, two test verification methods are presented:

1. Manual [semiautomated] using SQL Table Compare [Ref 34] functionality in TOAD®[24]
2. Automated using PL-SQL, HP Quick Test Professional® [test automation tool] and HP Quality Center® [testing management too]

Test Steps:

(1) Reload the data files into TestUser schema by Oracle Loader

```
DROP TABLE testuser.lt_storetransactions CASCADE CONSTRAINTS;
DROP DIRECTORY test_dir;
CREATE DIRECTORY test_dir AS 'C:\LANDING';
CREATE TABLE testuser.lt_storetransactions
( transactionrid VARCHAR2(50 BYTE),
  customerid VARCHAR2(50 BYTE),
  purchaseordernumber VARCHAR2(50 BYTE),
  orderdate VARCHAR2(50 BYTE),
  duedate VARCHAR2(50 BYTE),
  shipdate VARCHAR2(50 BYTE),
  order_status VARCHAR2(50 BYTE),
  onlineorderflag VARCHAR2(50 BYTE),
  salesid VARCHAR2(50 BYTE),
  storeid VARCHAR2(50 BYTE),
  invoice_num VARCHAR2(50 BYTE),
  billtoaddressid VARCHAR2(50 BYTE),
  shiptoaddressid VARCHAR2(50 BYTE),
  shipmethodid VARCHAR2(50 BYTE),
  currencyflag VARCHAR2(50 BYTE),
  currencyrateid VARCHAR2(50 BYTE),
  subtotal VARCHAR2(50 BYTE),
```

---

[24]   Registered Trade Marko of Quest Software Company

```
    order_taxamt VARCHAR2(50 BYTE),
    freight VARCHAR2(50 BYTE),
    comments VARCHAR2(50 BYTE),
    modifieddate VARCHAR2(50 BYTE)
)
ORGANIZATION EXTERNAL
   (  TYPE oracle_loader
      DEFAULT DIRECTORY test_dir
      ACCESS PARAMETERS
        (RECORDS DELIMITED BY NEWLINE
         STRING SIZES ARE IN CHARACTERS
         BADFILE test_dir:'lt_STORETRANSACTIONS.bad'
        LOGFILE test_dir:'lt_STORETRANSACTIONS.log'
         FIELDS TERMINATED BY ','
         OPTIONALLY ENCLOSED BY '"' AND '"'
         MISSING FIELD VALUES ARE NULL
         (transactionrid  CHAR(50),
          customerid  CHAR(50),
          purchaseordernumber  CHAR(50),
          orderdate  CHAR(50),
          duedate  CHAR(50),
          shipdate  CHAR(50),
          order_status  CHAR(50),
          onlineorderflag  CHAR(50),
          salesid  CHAR(50),
          storeid  CHAR(50),
          invoice_num  CHAR(50),
          billtoaddressid  CHAR(50),
          shiptoaddressid  CHAR(50),
          shipmethodid  CHAR(50),
          currencyflag  CHAR(50),
          currencyrateid  CHAR(50),
          subtotal  CHAR(50),
          order_taxamt  CHAR(50),
          freight  CHAR(50),
          comments  CHAR(50),
          modifieddate  CHAR(50)
         )
        )
      LOCATION (test_dir:'lt_STORETRANSACTIONS.DAT')
   )
REJECT LIMIT 0
NOPARALLEL
NOMONITORING;
```

(2) Copy Transaction data into test table with valid data format.

```
DROP TABLE testuser.t_storetransactions;
/
CREATE TABLE testuser.t_storetransactions
    (transactionrid,
     customerid,
     purchaseordernumber,
     orderdate,
     duedate,
     shipdate,
     order_status,
     onlineorderflag,
     salesid,
     storeid,
     invoice_num,
     billtoaddressid,
     shiptoaddressid,
     shipmethodid,
     currencyflag,
     currencyrateid,
     subtotal,
     order_taxamt,
     freight,
     comments,
     modifieddate
     ) AS
SELECT
    TRIM(transactionrid),
    TRIM(customerid),
    TO_NUMBER (TRIM(purchaseordernumber))
            AS purchaseordernumber,
    TO_DATE (TRIM(orderdate), 'YYYY-MM-DD')
            AS orderdate,
    TO_DATE (TRIM(duedate), 'YYYY-MM-DD')
            AS duedate,
    TO_DATE (TRIM(shipdate), 'YYYY-MM-DD')
            AS shipdate,
    TRIM(order_status),
    TRIM(onlineorderflag),
    TRIM(salesid),
    TRIM(storeid),
    TO_NUMBER (TRIM(invoice_num)) AS invoice_num,
    TRIM(billtoaddressid),
    TRIM(shiptoaddressid),
    TRIM(shipmethodid),
```

```
        TRIM(currencyflag),
        TRIM(currencyrateid),
        TO_NUMBER (TRIM(subtotal)) AS subtotal,
        TO_NUMBER (TRIM(order_taxamt)) AS order_taxamt,
        TRIM(freight),
        TRIM(comments),
        TO_DATE (TRIM(modifieddate), 'YYYY-MM-DD')
                AS modifieddate
    FROM testuser.lt_storetransactions;
```

(3) Create test result table in TESTUSER schema.

```
        DROP TABLE testuser.test_results;
        CREATE TABLE testuser.test_results
        ( item                VARCHAR2(50 BYTE),
          run_version         VARCHAR2(50 BYTE),
          runtime             DATE,
          total_failnmb       NUMBER,
          tablename_source    VARCHAR2(50 BYTE),
          tablename_target    VARCHAR2(50 BYTE),
          test_result         VARCHAR2(50 BYTE),
          comments            VARCHAR2(50 BYTE)
        )
        TABLESPACE users
        PCTUSED      0
        PCTFREE      10
        INITRANS     1
        MAXTRANS     255
        STORAGE      (INITIAL        64 k
                      MINEXTENTS      1
                      MAXEXTENTS       2147483645
                      PCTINCREASE     0
                      BUFFER_POOL     DEFAULT
                     )
        LOGGING
        NOCOMPRESS
        NOCACHE
        NOPARALLEL
        MONITORING;
```

(4) There are two testing options depending on the testing tools and management decision.

Testing Option I: Manually compare the data tables by using Toad.

(5) Compare Data by using Toad: Open Toad and use the compare data function on Database manual.

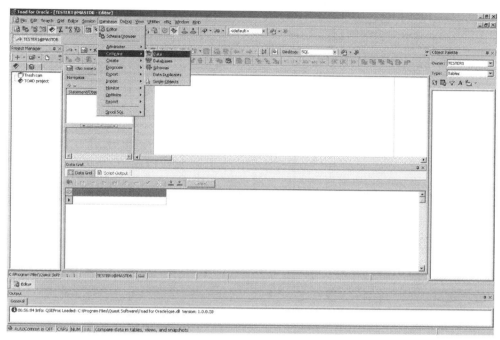

Figure 10.6

Select the table names to compare (STAGE.T_STORETRANSACTION and Table TESTUSER.T_STORETRANSACTION).

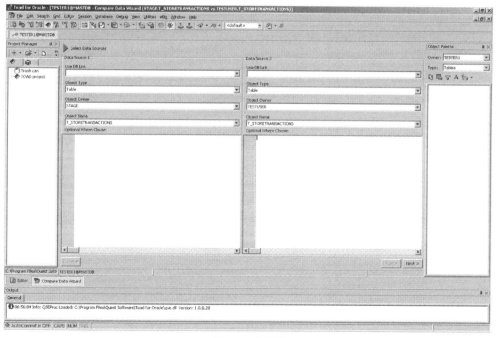

Figure 10.7

Keep the default setup.

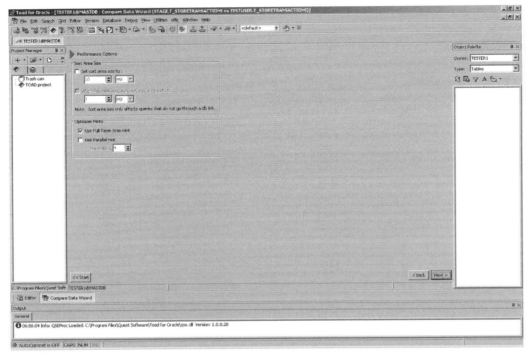

Figure 10.8

All the column names are displayed.

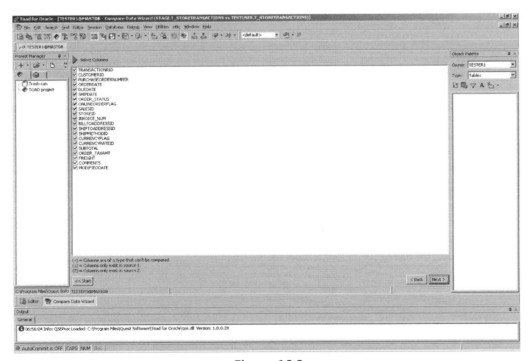

Figure 10.9

Make an order on the test result.

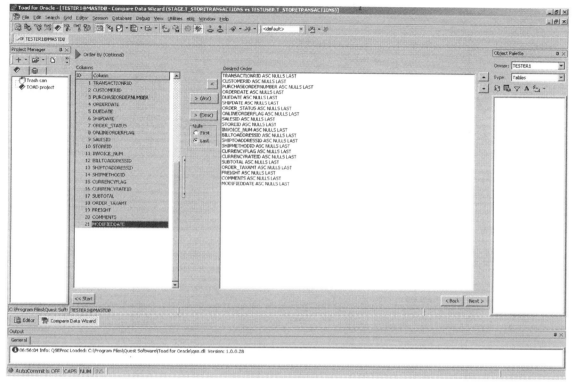

Figure 10.10

Result I: Rows in Table STAGE.T_STORETRANSACTION but not in Table TESTUSER.T_STORETRANSACTION.

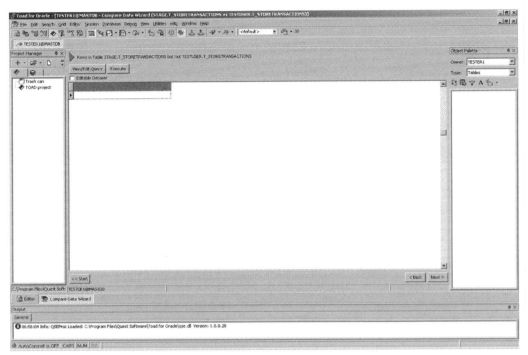

Figure 10.11

Tester also can see the query details by click on the "View/Edit Query" button.

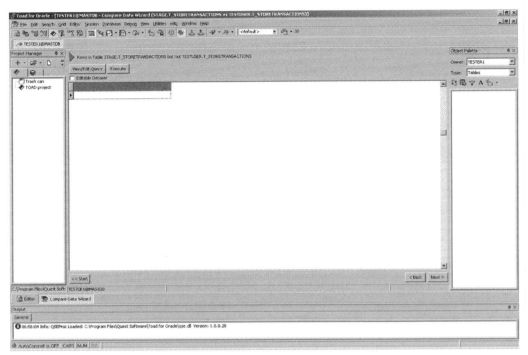

Figure 10.12

Result II: Rows in Table TESTUSER.T_STORETRANSACTION but not in Table STAGE.T_STORETRANSACTION.

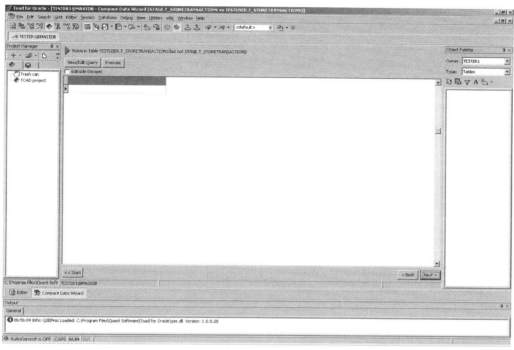

Figure 10.13

Here is the query detail again.

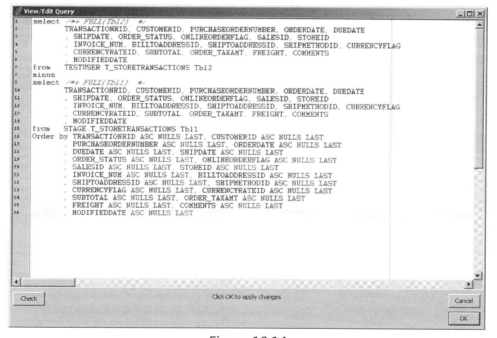

Figure 10.14

Result III: All the difference between the two tables.

Figure 10.15

The full SQL:

```
Select * from
  (
    select 'STAGE.T_STORETRANSACTIONS' "Row Source", a.* from
    (
select /*+ FULL(Tbl1) */
        TRANSACTIONRID, CUSTOMERID, PURCHASEORDERNUMBER, ORDERDATE, DUEDATE
        , SHIPDATE, ORDER_STATUS, ONLINEORDERFLAG, SALESID, STOREID
        , INVOICE_NUM, BILLTOADDRESSID, SHIPTOADDRESSID, SHIPMETHODID, CURRENCYFLAG
        , CURRENCYRATEID, SUBTOTAL, ORDER_TAXAMT, FREIGHT, COMMENTS
        , MODIFIEDDATE
from    STAGE.T_STORETRANSACTIONS Tbl1
minus
select /*+ FULL(Tbl2) */
        TRANSACTIONRID, CUSTOMERID, PURCHASEORDERNUMBER, ORDERDATE, DUEDATE
        , SHIPDATE, ORDER_STATUS, ONLINEORDERFLAG, SALESID, STOREID
        , INVOICE_NUM, BILLTOADDRESSID, SHIPTOADDRESSID, SHIPMETHODID, CURRENCYFLAG
        , CURRENCYRATEID, SUBTOTAL, ORDER_TAXAMT, FREIGHT, COMMENTS
        , MODIFIEDDATE
from    TESTUSER.T_STORETRANSACTIONS Tbl2
Order by TRANSACTIONRID ASC NULLS LAST, CUSTOMERID ASC NULLS LAST, PURCHASEORDERNUMBER ASC NULLS LAST, ORDERDATE ASC NULLS LAST
        , DUEDATE ASC NULLS LAST, SHIPDATE ASC NULLS LAST, ORDER_STATUS ASC NULLS LAST, ONLINEORDERFLAG ASC NULLS LAST
        , SALESID ASC NULLS LAST, STOREID ASC NULLS LAST, INVOICE_NUM ASC NULLS LAST, BILLTOADDRESSID ASC NULLS LAST
        , SHIPTOADDRESSID ASC NULLS LAST, SHIPMETHODID ASC NULLS LAST, CURRENCYFLAG ASC NULLS LAST, CURRENCYRATEID ASC NULLS LAST
        , SUBTOTAL ASC NULLS LAST, ORDER_TAXAMT ASC NULLS LAST, FREIGHT ASC NULLS LAST, COMMENTS ASC NULLS LAST, MODIFIEDDATE ASC NULLS LAST
      ) A
    union all
    select 'TESTUSER.T_STORETRANSACTIONS', b.* from
    (
select /*+ FULL(Tbl2) */
        TRANSACTIONRID, CUSTOMERID, PURCHASEORDERNUMBER, ORDERDATE, DUEDATE
        , SHIPDATE, ORDER_STATUS, ONLINEORDERFLAG, SALESID, STOREID
        , INVOICE_NUM, BILLTOADDRESSID, SHIPTOADDRESSID, SHIPMETHODID, CURRENCYFLAG
        , CURRENCYRATEID, SUBTOTAL, ORDER_TAXAMT, FREIGHT, COMMENTS
        , MODIFIEDDATE
from    TESTUSER.T_STORETRANSACTIONS Tbl2
minus
select /*+ FULL(Tbl1) */
        TRANSACTIONRID, CUSTOMERID, PURCHASEORDERNUMBER, ORDERDATE, DUEDATE
        , SHIPDATE, ORDER_STATUS, ONLINEORDERFLAG, SALESID, STOREID
        , INVOICE_NUM, BILLTOADDRESSID, SHIPTOADDRESSID, SHIPMETHODID, CURRENCYFLAG
        , CURRENCYRATEID, SUBTOTAL, ORDER_TAXAMT, FREIGHT, COMMENTS
        , MODIFIEDDATE
from    STAGE.T_STORETRANSACTIONS Tbl1
Order by TRANSACTIONRID ASC NULLS LAST, CUSTOMERID ASC NULLS LAST, PURCHASEORDERNUMBER ASC NULLS LAST, ORDERDATE ASC NULLS LAST
       , DUEDATE ASC NULLS LAST, SHIPDATE ASC NULLS LAST, ORDER_STATUS ASC NULLS LAST, ONLINEORDERFLAG ASC NULLS LAST
       , SALESID ASC NULLS LAST, STOREID ASC NULLS LAST, INVOICE_NUM ASC NULLS LAST, BILLTOADDRESSID ASC NULLS LAST
       , SHIPTOADDRESSID ASC NULLS LAST, SHIPMETHODID ASC NULLS LAST, CURRENCYFLAG ASC NULLS LAST, CURRENCYRATEID ASC NULLS LAST
       , SUBTOTAL ASC NULLS LAST, ORDER_TAXAMT ASC NULLS LAST, FREIGHT ASC NULLS LAST, COMMENTS ASC NULLS LAST, MODIFIEDDATE ASC NULLS LAST
      ) B
  )
Order by TRANSACTIONRID ASC NULLS LAST, CUSTOMERID ASC NULLS LAST, PURCHASEORDERNUMBER ASC NULLS LAST, ORDERDATE ASC NULLS LAST
       , DUEDATE ASC NULLS LAST, SHIPDATE ASC NULLS LAST, ORDER_STATUS ASC NULLS LAST, ONLINEORDERFLAG ASC NULLS LAST
       , SALESID ASC NULLS LAST, STOREID ASC NULLS LAST, INVOICE_NUM ASC NULLS LAST, BILLTOADDRESSID ASC NULLS LAST
       , SHIPTOADDRESSID ASC NULLS LAST, SHIPMETHODID ASC NULLS LAST, CURRENCYFLAG ASC NULLS LAST, CURRENCYRATEID ASC NULLS LAST
       , SUBTOTAL ASC NULLS LAST, ORDER_TAXAMT ASC NULLS LAST, FREIGHT ASC NULLS LAST, COMMENTS ASC NULLS LAST, MODIFIEDDATE ASC NULLS LAST, 1
```

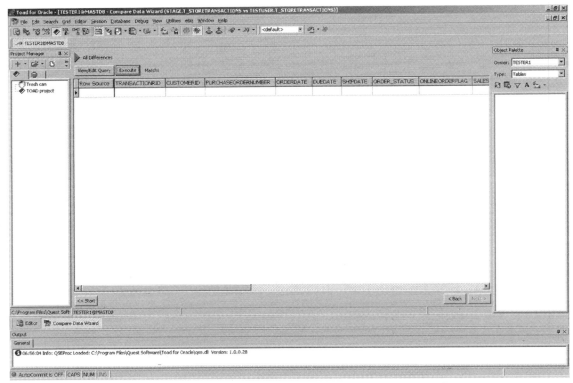

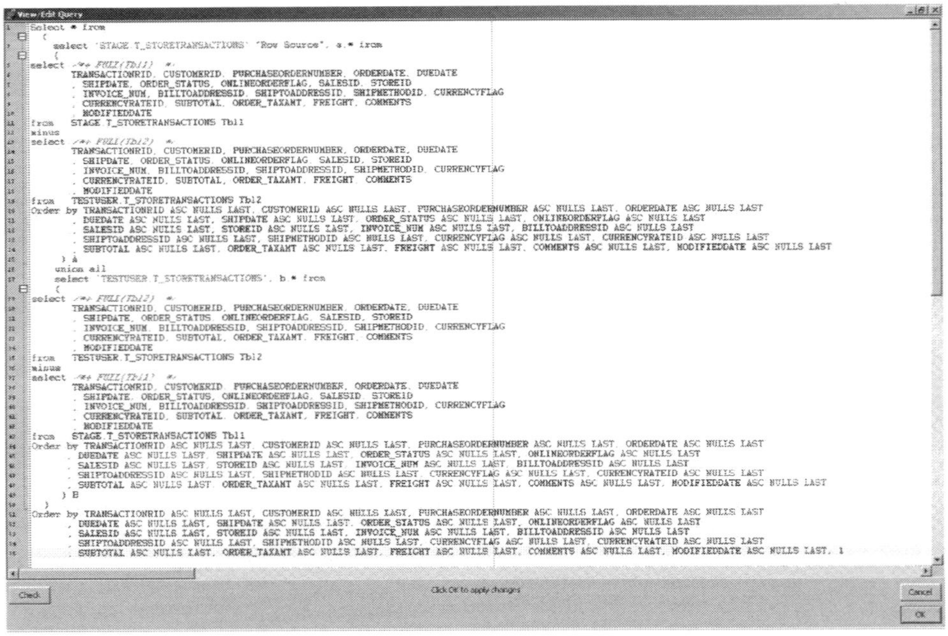

Figure 10.16

SQL below can be used for simple comparisons between two tables:

An example, comparing tables A and B, and the primary key of both tables is ID:

```
SELECT MIN(TableName) as TableName, ID, COL1,
        COL2, COL3 ...
FROM
( SELECT 'Table A' as TableName, A.ID, A.COL1,
          A.COL2, A.COL3, ...
    FROM A
    UNION ALL
  SELECT 'Table B' as TableName, B.ID, B.COL1,
          B.col2, B.COL3, ...
    FROM B
 ) tmp
GROUP BY ID, COL1, COL2, COL3 ...
HAVING COUNT(*) = 1
ORDER BY ID
```

Testing Option II: Automation testing script by using PL-SQL and HP Quick Test Professional; Testing Management Tool - HP Quality Center 10.

(5) Create generic test procedure and apply to testing database (Toad)

```
CREATE OR REPLACE Procedure TESTUSER.test002
(local_user IN VARCHAR2, target_user IN VARCHAR2, tablelike IN
VARCHAR2, Run_Version IN VARCHAR2)
IS
    tname VARCHAR2(30);
    CURSOR cur_table
    IS
        SELECT table_name
          FROM all_tables
         WHERE owner = local_user
         AND table_name LIKE tablelike;
    loc_table_name    VARCHAR2 (100);
    tar_table_name    VARCHAR2 (100);
    uatval            PLS_INTEGER;
    splval            PLS_INTEGER;
    sqlstatement      VARCHAR2 (2000);
    itemno PLS_INTEGER := 0;
BEGIN
    OPEN cur_table;
    LOOP
        FETCH cur_table INTO tname;
```

```
        EXIT WHEN cur_table%NOTFOUND;
        loc_table_name :=  local_user || '.' ||tname;
        tar_table_name :=  target_user || '.' ||tname;
        sqlstatement :=
'SELECT count(*) FROM (SELECT * FROM ' || loc_table_name || '
MINUS SELECT * FROM '|| tar_table_name || ')';
        EXECUTE IMMEDIATE sqlstatement
                    INTO uatval;
        sqlstatement :=
'SELECT count(*) FROM (SELECT * FROM ' || tar_table_name || ' MINUS
SELECT * FROM '|| loc_table_name || ')';
        EXECUTE IMMEDIATE sqlstatement
                    INTO sp1val;
        itemno := itemno + 1;
        IF uatval <> 0 OR sp1val <> 0
        THEN
        Insert into testuser.test_results
( Item, Run_Version, RunTime, Total_FailNMB, TableName_Source,
TableName_Target, Test_Result, Comments)
        Values ( itemno, Run_Version, SYSTIMESTAMP, uatval + sp1val,
loc_table_name, tar_table_name, 'Fail', itemno || '.' ||loc_table_
name || ' has different result!');
        ELSE
        Insert into testuser.test_results
( Item, Run_Version, RunTime, Total_FailNMB, TableName_Source,
TableName_Target, Test_Result, Comments)
Values ( itemno, Run_Version, SYSTIMESTAMP, uatval + sp1val, loc_
table_name, tar_table_name, 'Pass', null);
        END IF;
    END LOOP;
    CLOSE cur_table;
    COMMIT;
END;
```

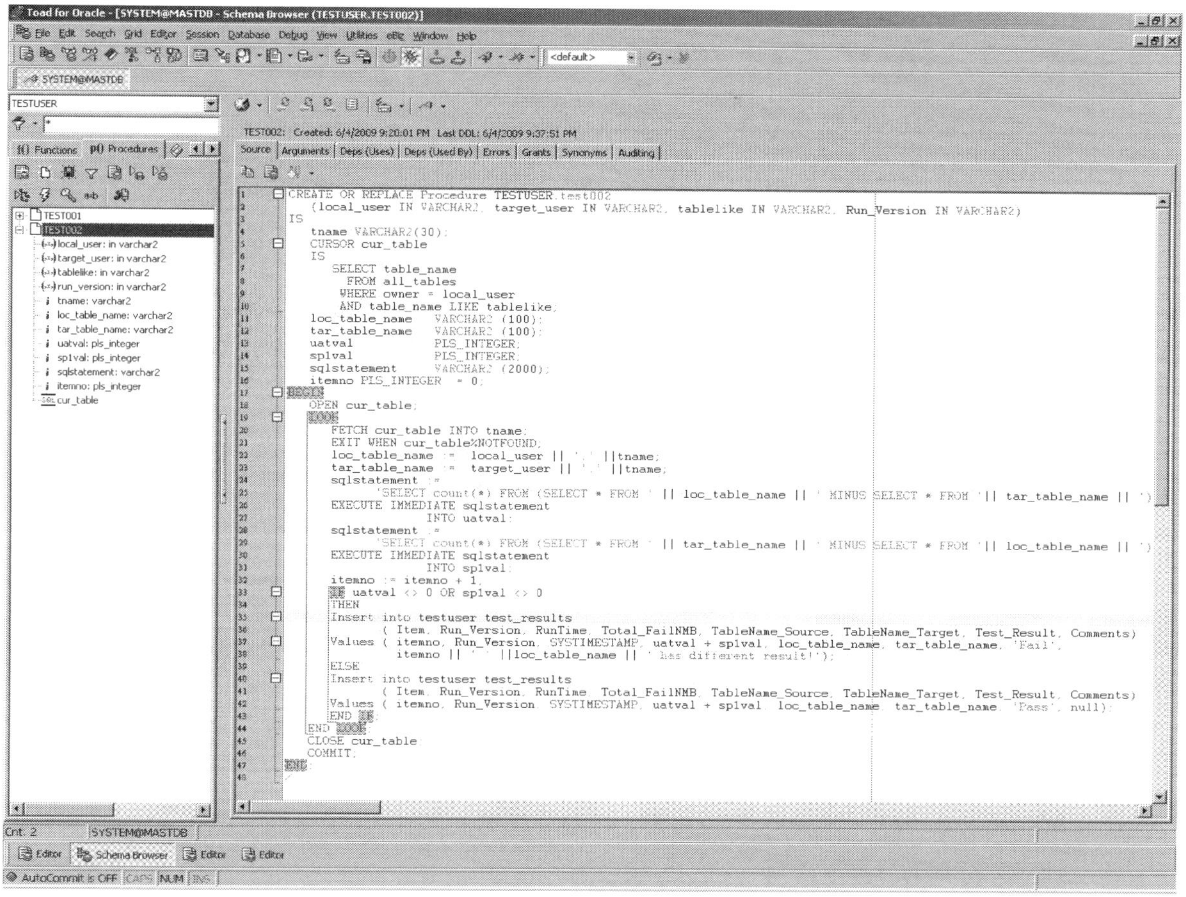

Figure 10.17: The test procedure is applied to database

(6) Execute the Test Procedure to compare the data in Staging database by HP Quick Test Professional 10 (VB Script).

```
P1= Parameter( "ProcedureName")
P2= Parameter( "Target")
P3= Parameter( "Actual")
P4= Parameter( "CheckPt")
P5= Parameter( "ReleaseNo")

RunStoredProcedure P1, P2, P3, P4, P5
VerifyTestResult  P5

Function RunStoredProcedure(ProcedureName, Parameter1, Parameter2,
Parameter3, Parameter4 )
        Dim Cmd
      DatabaseName="MASTDB"
      UserID="TESTUSER"
      Pwd="TEST1234"
```

```
        ' Create the database object
        Set Cmd = CreateObject("ADODB.Command")

        ' Activate the connection.
        Cmd.ActiveConnection = "DRIVER={Microsoft ODBC for
Oracle}; SERVER=" & DatabaseName & ";User ID=" & UserID &
";Password=" & Pwd & " ;"

        ' Set the command type to Stored Procedures
        Cmd.CommandType = 4
        Cmd.CommandText = ProcedureName

        ' Define Parameters for the stored procedure
        Cmd.Parameters.Refresh

        'Pass Parameters to Stored Procedure
        Cmd.Parameters(0).Value = Parameter1
        Cmd.Parameters(1).Value = Parameter2
        Cmd.Parameters(2).Value = Parameter3
        Cmd.Parameters(3).Value = Parameter4

        ' Execute the stored procedure
        Cmd.Execute()
        Set Cmd = Nothing
End Function

Function VerifyTestResult (Parameter4 )
        Dim Rs, Conn
        Dim SQL
    DatabaseName="MASTDB"
    UserID="TESTUSER"
    Pwd="TEST1234"
        ' Create the database object
        Set Conn=Createobject("adodb.connection")
        Set  Rs=Createobject("adodb.recordset")
        Conn.open="Driver={Microsoft ODBC for Oracle};Server= " &
DatabaseName &_
        ";Uid="& UserID &";Pwd="& Pwd &";"

        SQL="select TEST_RESULT, TABLENAME_SOURCE, COMMENTS  from
TESTUSER.TEST_RESULTS where RUN_VERSION = '"& Parameter4 &"' "
        Rs.open SQL,Conn,3,3

        Do While Not Rs.eof
                Results = Rs("TEST_RESULT")
                SourceTable = Rs("TABLENAME_SOURCE")
                If Results = "Pass" Then
                        Reporter.ReportEvent micPass, Parameter4
& "_Table Comparison", SourceTable
```

```
                 Else
                         Reporter.ReportEvent micfail,  Parameter4
    & "_Table Comparison", SourceTable
                 End if
                 Rs.movenext
         Loop
         Rs.close
         Set Rs=nothing
         Conn.close
         Set Conn=nothing
    End Function
```

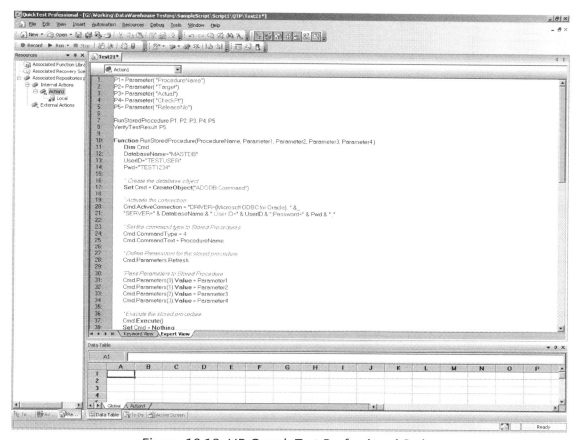

Figure 10.18: HP Quack Test Professional Script

(7) Run QTP script and save the test results into Quality Center

Note:

Before connecting HP Quick Test Professional to HP Quality Center, make sure the add-ins are installed in Quality Center:

- HP Quality Center Connectivity Add-in
- Quick Test Professional Add-in

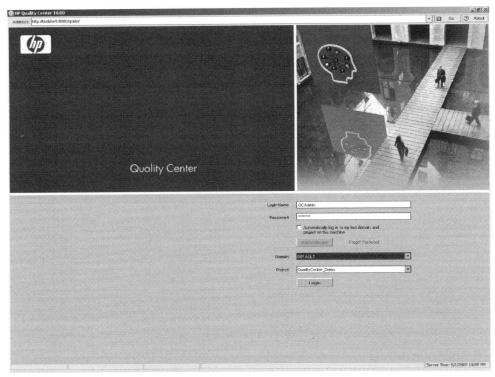

Figure 10.19: Execute the Quack Test Professional Script through HP Quality Center 10

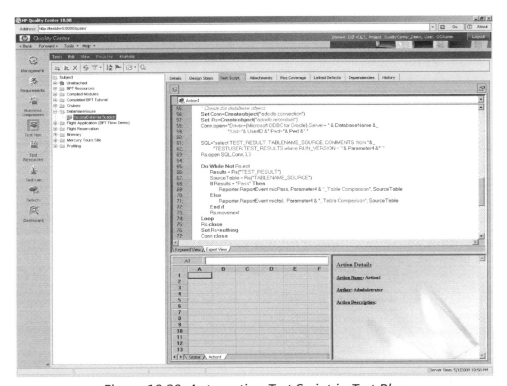

Figure 10.20: Automation Test Script in Test Plan

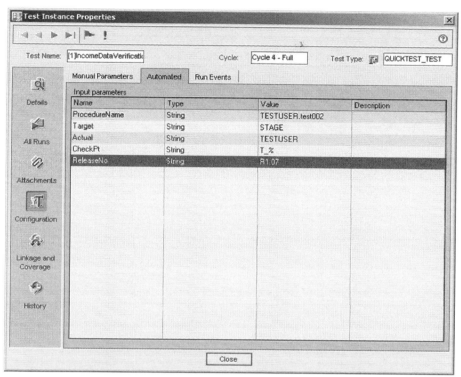

Figure 10.21: Update the script parameters in Quality Center

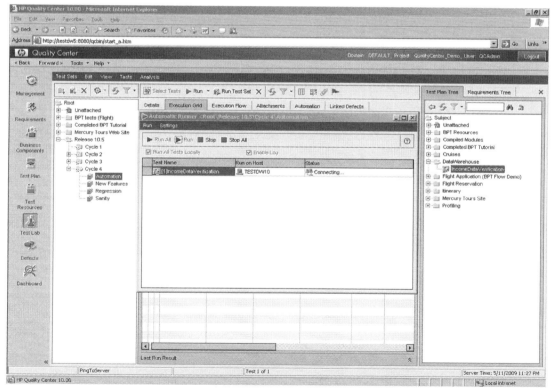

Figure 10.22: Run the script in Quality Center

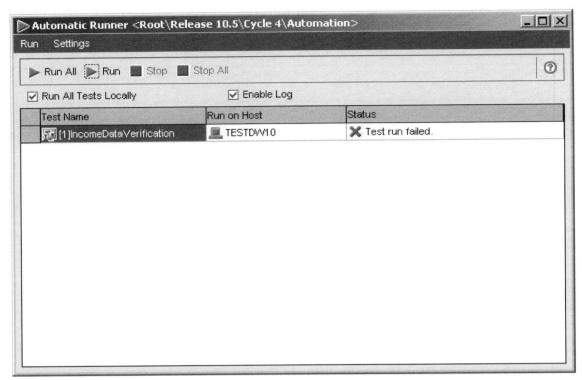

Figure 10.23: Test Result in Quality Center

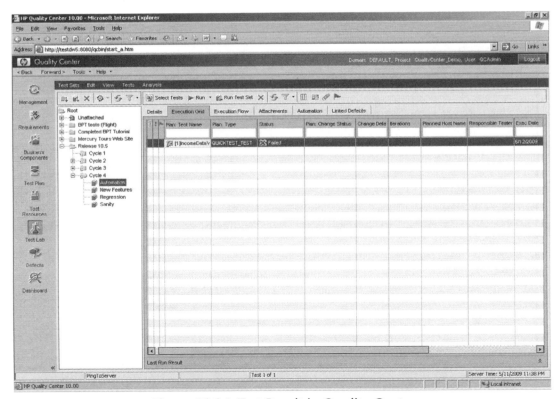

Figure 10.24: Test Result in Quality Center

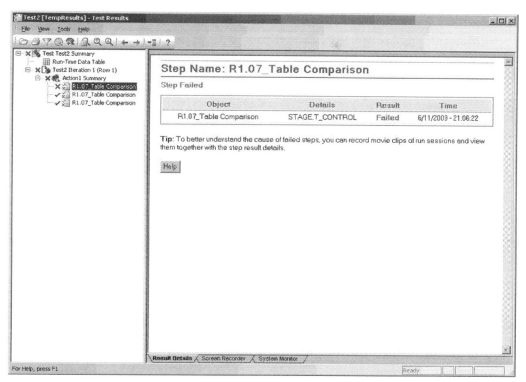

Figure 10.25: Test Result—Failed

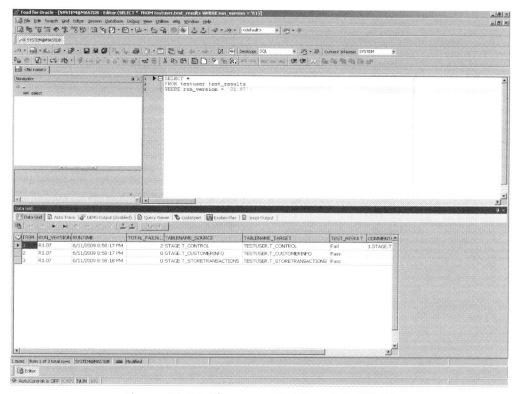

Figure 10.26: The same Test Result in TOAD

Test Case 2

<u>Scope and Description of Data Warehouse of Testing</u>: ETL verification [bases on business rules] in Master Database.

<u>Introduction</u>: There is an ETL job from Staging Database load the customer information data into data warehouse.

It includes some data format changes below.

	Column Name	Staging Database		Master Database	
		Format	Data Sample	Format	Data Sample
1	CUSTOMERID	Varchar2 (14byte)	CU1010000 00077	Number (10)	1000000077
2	CUSTOMERNAME	Different column for First and Last name	John	Smith	First and Last Name in a same column	John Smith
3	STATUS	None	Null	Active with Transactions	Active
4	CRDACCOUNTNUMBER	Varchar2 (20byte)	C10904000 0000016	Number (15)	1090400000 00016

Related Data Source: Oracle 10g Database

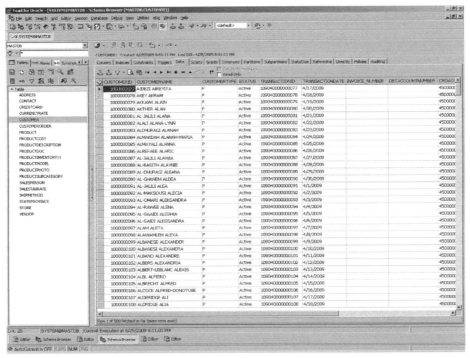

Figure 10.27: Customer Table in Master Database.

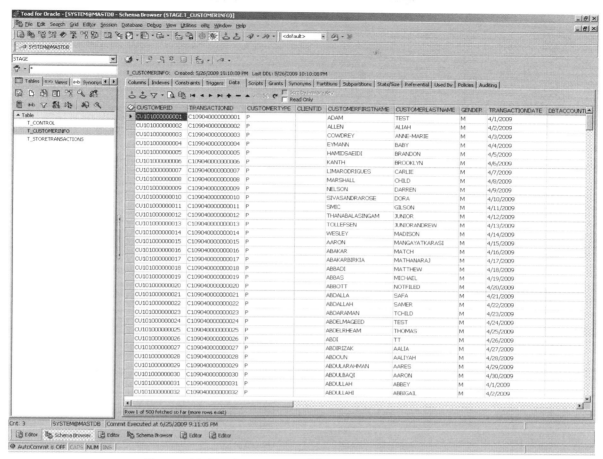

Figure 10.28: Customer Table in Staging Database.

Test Option I: Compare the data by using SQL script (Toad)

```
SELECT COUNT (*) AS Record_Count,
        DECODE(SUM(DECODE(mast.customerid, stag.customerid, 0 , 1)), 0,
'Pass', 'Fail') AS customerid,
        DECODE(SUM(DECODE(mast.customername, stag.customername, 0 ,
1)), 0, 'Pass', 'Fail') AS customername,
        DECODE(SUM(DECODE(mast.customertype, stag.customertype, 0 ,
1)), 0, 'Pass', 'Fail') AS customertype,
        DECODE(SUM(DECODE(mast.status, stag.status, 0 , 1)), 0, 'Pass',
'Fail') AS status,
        DECODE(SUM(DECODE(mast.transactionid, stag.transactionid, 0 ,
1)), 0, 'Pass', 'Fail') AS transactionid,
        DECODE(SUM(DECODE(mast.transactiondate, stag.transactiondate, 0
, 1)), 0, 'Pass', 'Fail') AS transactiondate,
        DECODE(SUM(DECODE(mast.invoice_number, stag.invoice_number, 0 ,
1)), 0, 'Pass', 'Fail') AS invoice_number,
```

```
        DECODE(SUM(DECODE(mast.dbtaccountnumber, stag.dbtaccountnumber,
0 , 1)), 0, 'Pass', 'Fail') AS dbtaccountnumber,
        DECODE(SUM(DECODE(mast.crdaccountnumber, stag.crdaccountnumber,
0 , 1)), 0, 'Pass', 'Fail') AS crdaccountnumber,
        DECODE(SUM(DECODE(mast.storeid, stag.storeid, 0 , 1)), 0,
'Pass', 'Fail') AS storeid,
        DECODE(SUM(DECODE(mast.addressid, stag.addressid, 0 , 1)), 0,
'Pass', 'Fail') AS addressid,
        DECODE(SUM(DECODE(mast.modifieddate, stag.modifieddate, 0 ,
1)), 0, 'Pass', 'Fail') AS modifieddate,
        DECODE(SUM(DECODE(mast.salesid, stag.salesid, 0 , 1)), 0, 'Pass',
'Fail') AS salesid
  FROM mastdb.customer mast,
        (SELECT TO_NUMBER (SUBSTR (customerid, -10)) AS customerid,
(customerfirstname || ' ' || customerlastname) AS customername,
                customertype,
                'Active' AS status,
                TO_NUMBER (SUBSTR (transactionid, -15)) AS
transactionid,
                transactiondate,
                NULL AS invoice_number,
                TO_CHAR (dbtaccountnumber) AS dbtaccountnumber,
                TO_CHAR (crdaccountnumber) AS crdaccountnumber,
                NULL AS storeid,
                addressid,
                modifieddate,
                NULL AS salesid
        FROM stage.t_customerinfo) stag
  WHERE mast.customerid = stag.customerid
AND mast.modifieddate = stag.modifieddate;
```

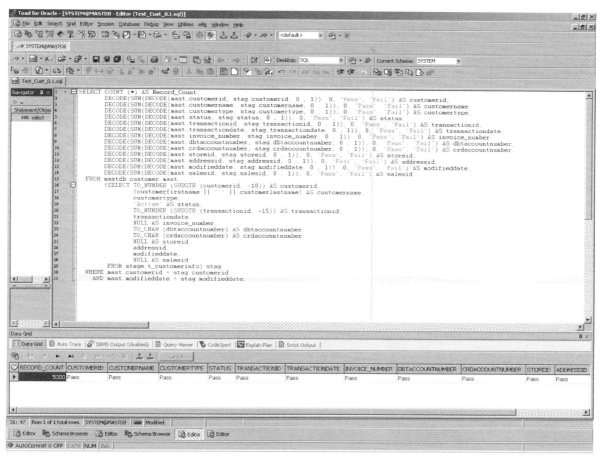

Figure 10.29: Test Script and results.

Test Option II: Compare the data by using SQL script insides VB script [Descriptive Programming]; the test result will be report to a Microsoft Excel data sheet.

The SQL Query:

```
SELECT COUNT (*) AS Record_Count,
       SUM(DECODE(mast.customerid, stag.customerid, 0 , 1)) AS
customerid,
       SUM(DECODE(mast.customername, stag.customername, 0 , 1)) AS
customername,
       SUM(DECODE(mast.customertype, stag.customertype, 0 , 1)) AS
customertype,
       SUM(DECODE(mast.status, stag.status, 0 , 1)) AS status,
       SUM(DECODE(mast.transactionid, stag.transactionid, 0 , 1)) AS
transactionid,
       SUM(DECODE(mast.transactiondate, stag.transactiondate, 0 , 1))
AS transactiondate,
```

```
        SUM(DECODE(mast.invoice_number, stag.invoice_number, 0 , 1)) AS
invoice_number,
        SUM(DECODE(mast.dbtaccountnumber, stag.dbtaccountnumber, 0 ,
1)) AS dbtaccountnumber,
        SUM(DECODE(mast.crdaccountnumber, stag.crdaccountnumber, 0 ,
1)) AS crdaccountnumber,
        SUM(DECODE(mast.storeid, stag.storeid, 0 , 1)) AS storeid,
        SUM(DECODE(mast.addressid, stag.addressid, 0 , 1)) AS
addressid,
        SUM(DECODE(mast.modifieddate, stag.modifieddate, 0 , 1)) AS
modifieddate,
        SUM(DECODE(mast.salesid, stag.salesid, 0 , 1)) AS salesid
  FROM mastdb.customer mast,
        (SELECT TO_NUMBER (SUBSTR (customerid, -10)) AS customerid,
        (customerfirstname || ' ' || customerlastname) AS
customername,
                customertype,
                'Active' AS status,
                TO_NUMBER (SUBSTR (transactionid, -15)) AS
transactionid,
                transactiondate,
                NULL AS invoice_number,
        TO_CHAR (dbtaccountnumber) AS dbtaccountnumber,
        TO_CHAR (crdaccountnumber) AS crdaccountnumber,
                NULL AS storeid,
                addressid,
                modifieddate,
                NULL AS salesid
        FROM stage.t_customerinfo) stag
  WHERE mast.customerid = stag.customerid
   AND mast.modifieddate = stag.modifieddate;
```

The details of VB Script:

```
'=====================================================
' VBScript Source File -- Database Testing
' Version: 1.0
' Feature and Description: ETL Verification
' Author:
' Date:
'=====================================================
Dim DBName, UserID, Pwd, TableName, SQL
Dim objXL, objWb, objP1, objP2 ' Excel object variables

' Open Excel data sheet
```

```
Set objXL = CreateObject ("Excel.Application")
objXL.Visible = true ' Show Excel window
objXL.DisplayAlerts = False 'Disable Excel Alert
Set objWb = objXL.WorkBooks.Open(GetPath + "DBTest.xls")
Set objP1 = objXL.ActiveWorkBook.WorkSheets("General_Info")
Set objP2 = objXL.ActiveWorkBook.WorkSheets("Test_Results")
objP1.Activate   ' Show General_Info Page

' Retrieve Database configuration information from the Excel file
DBName = Trim(objP1.Cells(12, 3).Value)
UserID = Trim(objP1.Cells(13, 3).Value)
Pwd = Trim(objP1.Cells(14, 3).Value)
TableName = Trim(objP1.Cells(15, 3).Value)
SQL = Trim(objP1.Cells(16, 3).Value)
SaveAsName = Trim(objP1.Cells(8, 3).Value)

' Apply default value
If CStr(SaveAsName) = "" Then
    SaveAsName = "Test_Results"
End If

' Manually enter User ID
If UserID = "" Then
    UserID = Inputbox ("Please enter your Database User ID:")
    UserID = Trim (UserID)
End If

' Manually enter Password
If Pwd = "" Then
    Pwd = Inputbox ("Please enter your Database Password:")
    Pwd = Trim (Pwd)
End If

objP2.Activate   ' Show Test_Results Page
' Create the database object and connect to database
Dim oCon: Set oCon = WScript.CreateObject("ADODB.Connection")
Dim oRs: Set oRs = WScript.CreateObject("ADODB.Recordset")
Dim ConnectInf
If oCon is Nothing Then
      Msgbox "Database Connection Error"
      Wscript.Quit
End If
ConnectInf = "dsn=" & DBName & "; uid="& UserID &"; pwd="& Pwd
&";"
oCon.Open ConnectInf
'   Execute the query
```

```
Set oRs = oCon.Execute(SQL)
If Err.number <> 0 Then
      Msgbox "SQL Error!!"
      Wscript.Quit
End If


' Read all query records and insert the results into excel file
Do until oRs.EOF
   For i = 0 to oRs.Fields.Count-1 'Get the Field Count
       objP2.Cells(i+2, 1).Value = "'" & CStr(i+1)
       objP2.Cells(i+2, 2).Value = "'" & TableName
       'Get Field Name
       objP2.Cells(i+2, 3).Value = "'" & oRs.Fields(i).Name
       'Get Field Value
       objP2.Cells(i+2, 4).Value = "'" & oRs.Fields(i).Value
       ' Check the test results without the row of record count
  If UCase(oRs.Fields(i).Name) <> "RECORD_COUNT" Then
          If oRs.Fields(i).Value = 0 Then
             ' There is no different record
             objP2.Cells(i+2, 5).Value = "'Pass"
          Else
             ' There is different record
             objP2.Cells(i+2, 5).Value = "'Fail"
             objP2.Cells(i+2, 5).Font.ColorIndex = 3 'Make color red
             objP2.Cells(i+2, 5).Font.Bold = True 'Make font bold
          End If
       End If
   Next
   oRs.MoveNext
Loop

oRs.Close
oCon.Close
Set oRs = Nothing
Set oCon = Nothing

' Save results
objWb.SaveAs GetPath + SaveAsName + ".xls"

objXL.DisplayAlerts = True
objXL.Quit
Set objP1 = Nothing
Set objP2 = Nothing
Set objWb = Nothing
Set objXL = Nothing
```

```
Function GetPath()
    ' Retrieve the script path
    DIM path
    path = WScript.ScriptFullName ' Script name
    GetPath = Left(path, InstrRev(path, "\"))
End Function
```

The test steps:

Step 1: Create an Excel test template following the format below:

Page 1—"General_Info" and insert the Query inside the SQL cell:

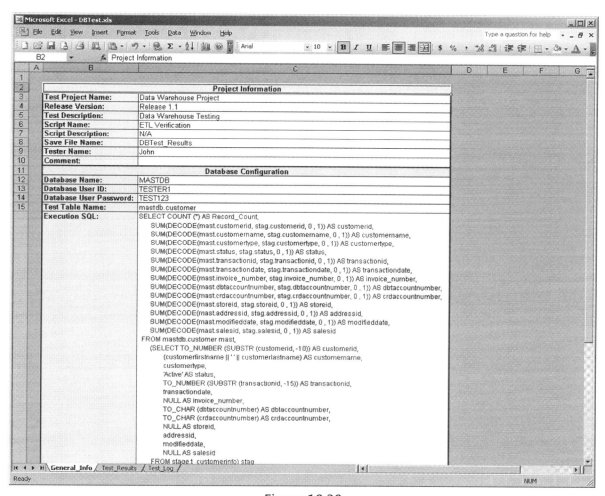

Figure 10.30

Data Sheet Page 2—"Test_Results":

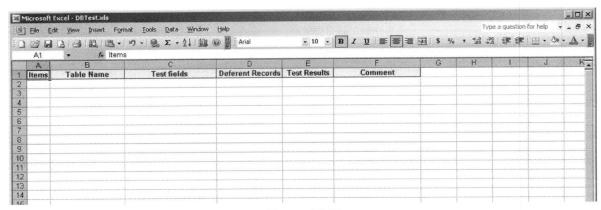

Figure 10.31

Step 2: Copy the VB Script into a Notepad and save the file with an extension name "vbs." Make sure this vbs file and the Excel template are located in a same folder. Here our example file name is Test_Script_V1.0.vbs and DBTest.xls.

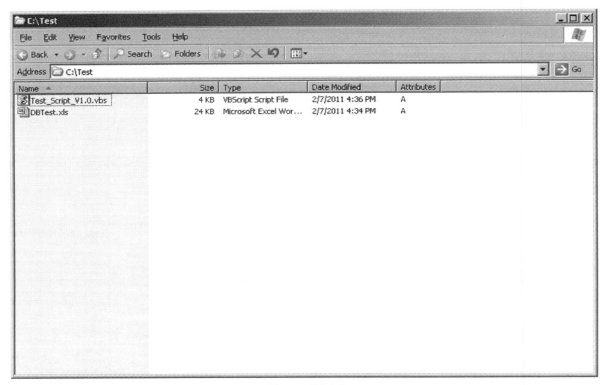

Figure 10.32

Step 3: Double-click on the Test_Script_V1.0.vbs file to run it. The test result file is generated.

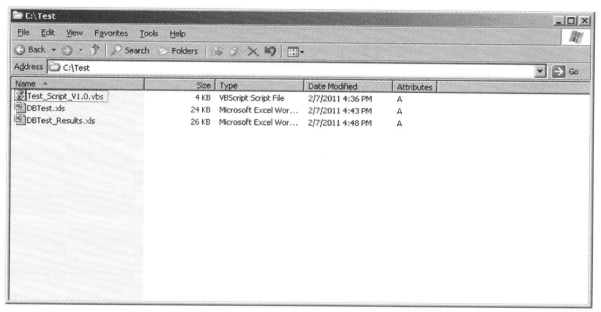

Figure 10.33

Step 4: Check the execution report on the result file.

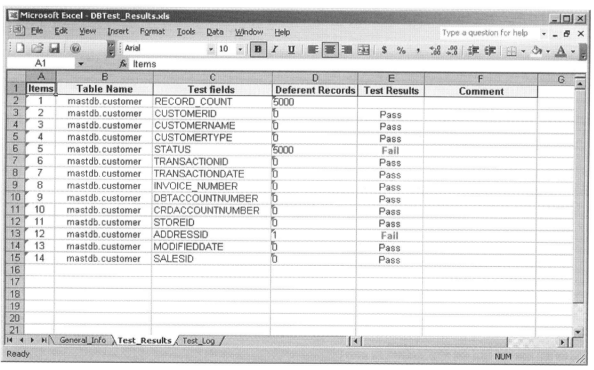

Figure 10.34

Test Case 3

<u>Scope and Description of Data Warehouse of Testing</u>: Landing files verification by using UNIX Shell Script.

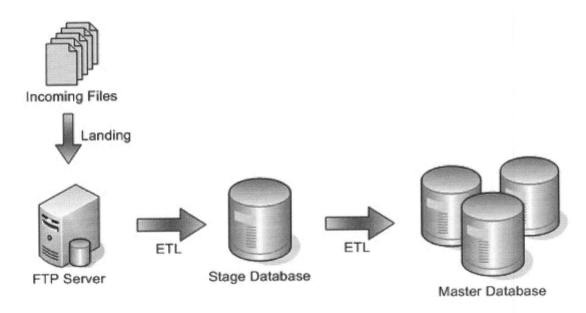

Figure 10.35

There are two files will be loaded into landing area (FTP server) every day. Here are the file names and directory:

FTP directory: /var/ftp/Landing/
Header/Trailer file: transaction.ht

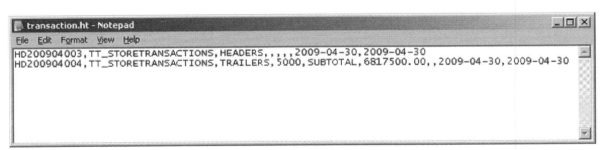

Figure 10.36: Header/Trailer file

Data file: customerinfo.csv

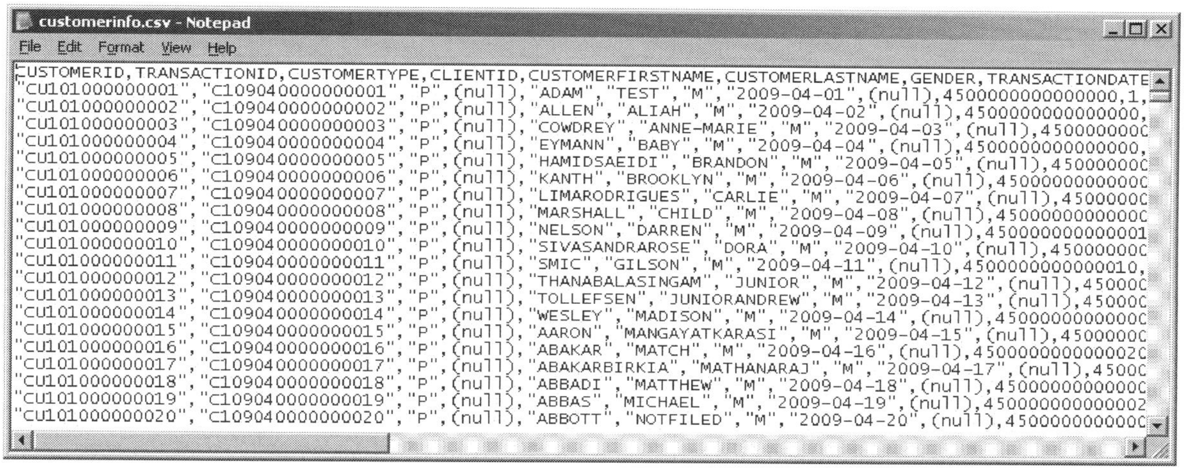

```
customerinfo.csv - Notepad
File Edit Format View Help
CUSTOMERID,TRANSACTIONID,CUSTOMERTYPE,CLIENTID,CUSTOMERFIRSTNAME,CUSTOMERLASTNAME,GENDER,TRANSACTIONDATE
"CU101000000001","C109040000000001","P",(null),"ADAM","TEST","M","2009-04-01",(null),4500000000000000,1,
"CU101000000002","C109040000000002","P",(null),"ALLEN","ALIAH","M","2009-04-02",(null),4500000000000000,
"CU101000000003","C109040000000003","P",(null),"COWDREY","ANNE-MARIE","M","2009-04-03",(null),450000000
"CU101000000004","C109040000000004","P",(null),"EYMANN","BABY","M","2009-04-04",(null),4500000000000000,
"CU101000000005","C109040000000005","P",(null),"HAMIDSAEIDI","BRANDON","M","2009-04-05",(null),45000000
"CU101000000006","C109040000000006","P",(null),"KANTH","BROOKLYN","M","2009-04-06",(null),4500000000000
"CU101000000007","C109040000000007","P",(null),"LIMARODRIGUES","CARLIE","M","2009-04-07",(null),4500000
"CU101000000008","C109040000000008","P",(null),"MARSHALL","CHILD","M","2009-04-08",(null),4500000000000
"CU101000000009","C109040000000009","P",(null),"NELSON","DARREN","M","2009-04-09",(null),4500000000000001
"CU101000000010","C109040000000010","P",(null),"SIVASANDRAROSE","DORA","M","2009-04-10",(null),45000000
"CU101000000011","C109040000000011","P",(null),"SMIC","GILSON","M","2009-04-11",(null),4500000000000010,
"CU101000000012","C109040000000012","P",(null),"THANABALASINGAM","JUNIOR","M","2009-04-12",(null),45000
"CU101000000013","C109040000000013","P",(null),"TOLLEFSEN","JUNIORANDREW","M","2009-04-13",(null),45000
"CU101000000014","C109040000000014","P",(null),"WESLEY","MADISON","M","2009-04-14",(null),4500000000000
"CU101000000015","C109040000000015","P",(null),"AARON","MANGAYATKARASI","M","2009-04-15",(null),4500000
"CU101000000016","C109040000000016","P",(null),"ABAKAR","MATCH","M","2009-04-16",(null),45000000000000020
"CU101000000017","C109040000000017","P",(null),"ABAKARBIRKIA","MATHANARAJ","M","2009-04-17",(null),45000
"CU101000000018","C109040000000018","P",(null),"ABBADI","MATTHEW","M","2009-04-18",(null),4500000000000
"CU101000000019","C109040000000019","P",(null),"ABBAS","MICHAEL","M","2009-04-19",(null),4500000000000002
"CU101000000020","C109040000000020","P",(null),"ABBOTT","NOTFILED","M","2009-04-20",(null),4500000000000
```

Figure 10.37: Data file—customerinfo.csv

This Bash Shell script will check the file existence, the loading date, and the data creation date.

```
#!/bin/bash
# This script gives information about a file.

LoadDate="$(stat -c %x /var/ftp/Landing/transaction.ht | awk
'{print $1}')"
DateInFile="$(cut -f8 -d"," /var/ftp/Landing/customerinfo.csv | head
-n 1)"
TodayDate="$(date +%Y-%m-%d)"
YesterdayDate="$(date -d yesterday +%Y-%m-%d)"

# Check the file existence
if [ -f /var/ftp/Landing/transaction.ht ]; then
   echo "Header and Trailer file not exist..."
   exit 1
fi
if [ -f /var/ftp/Landing/customerinfo.csv ]; then
   echo "Transaction data not exist..."
   exit 1
fi

#Check the file landing date
if [ "$LoadDate" == "$TodayDate" ]
then
echo "The file is loaded today."
else
echo "The file is NOT loaded today."
# Mail to tester
```

```
mail -s "The loading date is not match..." tester@abc.com
fi

#Check the date inside file
if [ "$DateInFile" == "$YesterdayDate" ]
then
echo "The data is generated yesterday."
else
echo "The data is NOT generated yesterday."
# Mail to tester
mail -s "The creation date is not match..." tester@abc.com
fi
```

The shell execution:

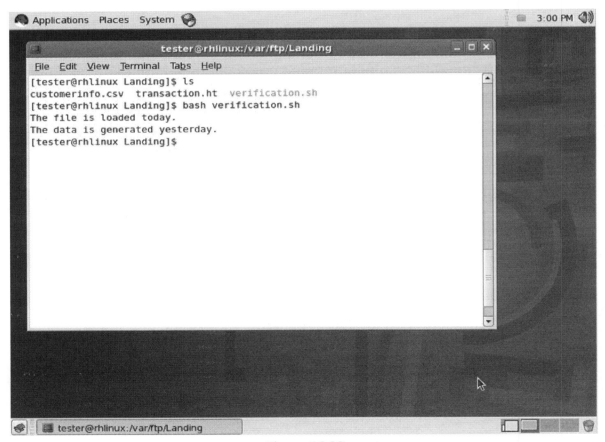

Figure 10.38

Test Case #4

Scope and Description of Data Warehouse of Testing:
Report Testing (VBScript Automation)

Testing Background:
Business requires a Daily Sales Report to summary the last day's sale information for the stores. Test team needs to create an automation script to compare the report data in data warehouse.

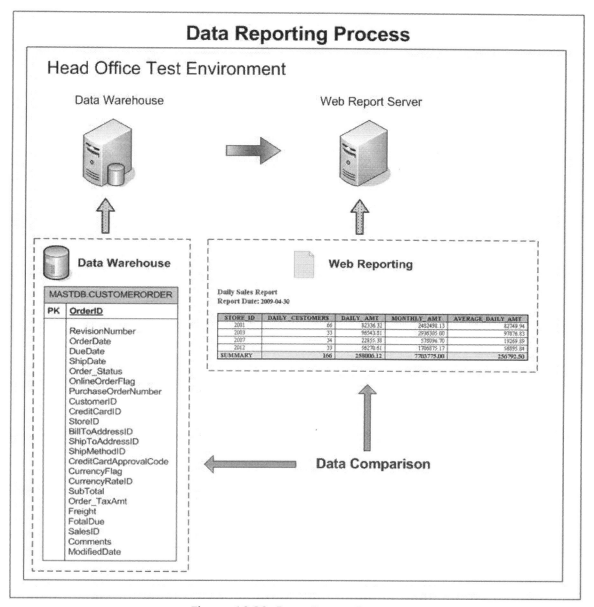

Figure 10.39: Data Report Process

The data in the "mastdb.customerorder" table:

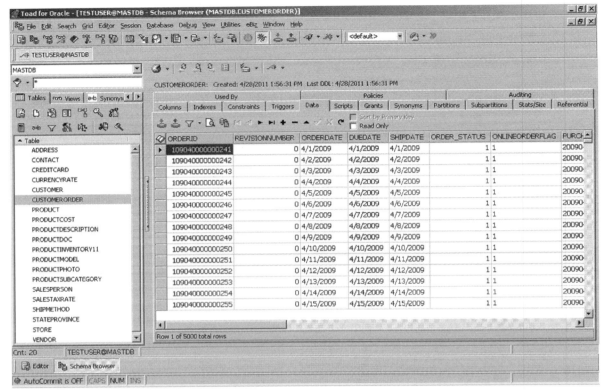

Figure 10.40

Sample Report: Daily Sales Report

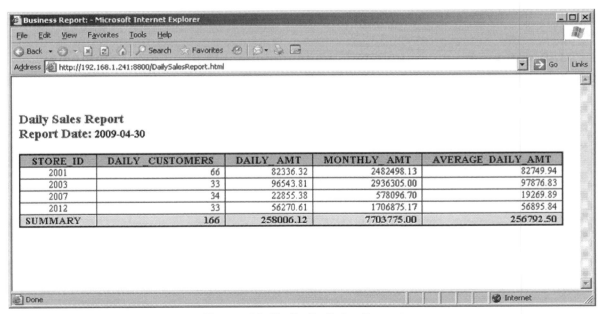

Figure 10.41: Daily Sales Report

The Automation Testing Script (VB Script):

```
Option Explicit
Public objIE 'Create object of IE Application
Dim OutputText

'Get the database access information
Dim SID, DBName, UserID, Password
SID = "MASTDB"
DBName = "MASTDB"

'Enter the output file name
Dim ConnectInf, Query
Dim OutputInfo, OutputFile
OutputFile = "TestResults.txt"
OutputInfo = ""

'Enter Database User ID
UserID = Inputbox ("Please enter your User ID:")
UserID = Trim (UserID)

'Enter Database Password
Password = Inputbox ("Please enter your Password:")
Password = Trim (Password)

'----------- Open IE -------------
'Navigate the report webpage.
NavWeb("http://192.168.1.241:8800/")
'Click the link: Daily Sales Report
WebLink "Daily Sales Report"

'Create the database connection
Dim oCon: Set oCon = WScript.CreateObject("ADODB.Connection")
Public oRs: Set oRs = WScript.CreateObject("ADODB.Recordset")

If oCon is Nothing then
   Msgbox "Connection error!"
End if

ConnectInf = "Provider=OraOLEDB.Oracle;Data Source="&_
             SID"; User ID="& UserID &"; Password="&_
             Password &";"
oCon.Open ConnectInf

'Create the testing output file with a time stamp.
```

```
CreateOutput OutputFile, vbCrLf & Cstr(Now) & " ------------------" &
vbCrLf

'---------------- Query I ----------------
'Get the last date from database.
Query = "SELECT TO_CHAR (MAX (orderdate), 'YYYY-MM-DD') AS orderdate
FROM mastdb.customerorder"
'Execute the query
SQLExecution Query

'Compare the Report Date value on the report web page.
If oRs.Fields(0).Value = ReadInfo("Report Date:", 10) Then
    OutputInfo = "The value of Report Date is correct!"
Else
    OutputInfo = "Database has a different report date:"
                    & oRs.Fields(0).Value
End If

'Write the result into test output file.
CreateOutput OutputFile, OutputInfo

'---------------- Query II ----------------
'Get the daily summery from database.
Query = "SELECT storeid AS store_id, COUNT(customerid) AS daily_
customers, SUM (totalpurchase) AS daily_amt  FROM (SELECT customerid,
storeid, SUM (subtotal + order_taxamt) AS totalpurchase FROM mastdb.
customerorder WHERE orderdate = (SELECT MAX(orderdate) FROM mastdb.
customerorder) GROUP BY storeid, customerid) GROUP BY storeid ORDER
BY storeid"

'Execute the query
SQLExecution Query

Dim ColumnNbr, RowNbr
Dim oF
Dim TotalCustomers, TotalDailyAmt, TotalMonthlyAmt, AvgDailyAmt
OutputInfo = ""
RowNbr = 1
TotalCustomers = 0
TotalDailyAmt = 0
TotalMonthlyAmt = 0
AvgDailyAmt = 0
OutputInfo = ""

'Compare the value on the report web page.
```

```
'This part will check the first three columns: STORE_ID, DAILY
_CUSTOMERS, and DAILY_ AMT
'Loop for different rows
While Not oRs.EOF
    ColumnNbr = 0
    'Loop for different columns
    For Each oF in oRs.Fields
        'Compare the query result and the value on web report
        If Round(oRs.Fields(ColumnNbr).Value, 2) <>
CDbl(WebTable(RowNbr+1, ColumnNbr+1)) Then
            OutputInfo = OutputInfo & "The database has a different
value for cell (" & CStr(RowNbr + 1) & ", " & CStr(ColumnNbr + 1)  &
"): "oRs.Fields(ColumnNbr).Value & vbCrLf
        End If
        Select Case ColumnNbr
            'Sum the DAILY_CUSTOMERS Summary
            Case 1
                TotalCustomers = TotalCustomers + Round(oRs.
Fields(ColumnNbr).Value, 2)
            'Sum the DAILY_AMT Summary
            Case 2
                TotalDailyAmt = TotalDailyAmt + Round(oRs.Fields(ColumnNbr).
Value, 3)
        End Select
        ColumnNbr = ColumnNbr + 1
    Next
    oRs.MoveNext
    RowNbr = RowNbr + 1
Wend


'Compare the query result and the value on web report for DAILY_
CUSTOMERS Summary
If TotalCustomers <> CDbl(WebTable(RowNbr + 1, 2)) Then
    OutputInfo=OutputInfo & "The database has a different DAILY_
CUSTOMERS Summary value: " & TotalCustomers & vbCrLf
End If
'Compare the query result and the value on web report for DAILY_AMT
Summary
If TotalDailyAmt <> CDbl(WebTable(RowNbr + 1, 3)) Then
    OutputInfo = OutputInfo & "The database has a different DAILY_AMT
Summary value: "TotalDailyAmt & vbCrLf
End If


"--------------- Query III ----------------
"Check how many days in the current month
```

```
Query = "SELECT COUNT (1) FROM (SELECT mcoo.orderdate FROM (SELECT *
FROM mastdb.customerorder mco, (SELECT MAX(orderdate) AS maxorderdate
FROM mastdb.customerorder) ldt WHERE TO_CHAR (mco.orderdate, 'Year')
= TO_CHAR (ldt.maxorderdate, 'Year') AND TO_CHAR (mco.orderdate,
'Month') = TO_CHAR (ldt.maxorderdate, 'Month')) mcoo GROUP BY mcoo.
orderdate)"

'Execute the query
SQLExecution Query

Dim DateNbr
DateNbr = CInt(oRs.Fields(0).Value)

''---------------- Query IV ----------------
''Get the Monthly Amount
Query = "Select mcoo.storeid, SUM (mcoo.subtotal + mcoo.order_taxamt)
AS totalpurchase FROM (SELECT * FROM mastdb.customerorder mco,
(SELECT MAX (orderdate) AS maxorderdate FROM mastdb.customerorder)
ldt WHERE TO_CHAR (mco.orderdate, 'Year') = TO_CHAR (ldt.
maxorderdate, 'Year') AND TO_CHAR (mco.orderdate, 'Month') = TO_CHAR
(ldt.maxorderdate, 'Month')) mcoo GROUP BY mcoo.storeid  ORDER BY
storeid"

'Execute the query
SQLExecution Query

Dim AverageAmt
RowNbr = 1

'Compare the value on the report web page.
'This part will check the last two columns: MONTHLY_AMT and
AVERAGE_DAILY_AMT
'Loop for different rows
While Not oRs.EOF
    If Round(oRs.Fields(1).Value, 2) <> CDbl(WebTable(RowNbr + 1, 4))
Then
    OutputInfo=OutputInfo & "The database has a different value for
cell (" & CStr(RowNbr+1) & ", 4): " & oRs.Fields(1).Value & vbCrLf
    End If
    AverageAmt = Round((CDbl(oRs.Fields(1).Value)/DateNbr), 2)

    If AverageAmt <> CDbl(WebTable(RowNbr + 1, 5)) Then
    OutputInfo=OutputInfo & "The database has a different value for
cell (" & CStr(RowNbr+1) &", 5): " & CStr(AverageAmt) & vbCrLf
    End If
```

```
        TotalMonthlyAmt = TotalMonthlyAmt + Round(oRs.Fields(1).Value, 2)

    oRs.MoveNext
    RowNbr = RowNbr + 1
Wend

'Compare the query result and the value on web report for MONTHLY_
AMT Summary
If TotalMonthlyAmt <> CDbl(WebTable(RowNbr + 1, 4)) Then
    OutputInfo = OutputInfo & "The database has a different MONTHLY_
AMT Summary value: " & TotalCustomers & vbCrLf
End If

'Compare the query result and the value on web report for AVERAGE_
DAILY_AMT Summary
AvgDailyAmt = Round((TotalMonthlyAmt / DateNbr), 2)
If AvgDailyAmt <> CDbl(WebTable(RowNbr + 1, 5)) Then
    OutputInfo=OutputInfo & "The database has a different AVERAGE_
DAILY_AMT Summary value: " & AvgDailyAmt & vbCrLf
End If

'Problem output
If OutputInfo = "" Then
    CreateOutput OutputFile, "The entire test data is correct!"
Else
    CreateOutput OutputFile, "Failed on the Data Verification: " &
vbCrLf & OutputInfo
End If

Msgbox "Script Finish!"

Err.Clear
Set oRs = Nothing
Set oCon = Nothing
Set objIE = Nothing

'================= Functions =================
'SQL Query Execution
Function SQLExecution(SQL)
    'SQL is the Query details
    Set oRs = oCon.Execute(Query)
    If Err.number <> 0 Then
        Msgbox "SQL Query Failed !"
        Wscript.Quit
    End If
    If oRs.EOF Then
```

```
        Msgbox "No query result!"
        Wscript.Quit
    End If
End Function


'Read infomation from the innerText on web page.
Function ReadInfo(ReadBefore, ReadSize)
   'ReadBefore is the Text before reading area.
   'ReadSize is the reading character number.
   'Sample: ReadInfo("Report Date:", 10)
    Dim FullText
    Dim StartPort
    FullText = objIE.Document.body.innerText
    StartPort = (InStr(LCase(FullText), LCase(ReadBefore)) +
Len(ReadBefore))
    ReadInfo = Trim(Right(FullText, Len(FullText) - StartPort))
    ReadInfo = Left(ReadInfo, ReadSize)
End Function


'Read infomation from the table on web page.
Function WebTable(RowID, ColumnID)
   'RowID is the row number on the table.
   'ColumnID is the column number on the table.
   'Sample: WebTable(2, 1)
    Dim Table
    Table = objIE.Document.getElementsByTagName("table")
    WebTable = Table.rows(RowID - 1).cells(ColumnID - 1).innerText
End Function


'Check the IE activity status
Function IEActive()
    Do While objIE.Busy Or (objIE.READYSTATE <> 4)
        Wscript.Sleep 80
    Loop
    WScript.Sleep 80
End Function


'Navigate Website
Function NavWeb(WebAddress)
    Set objIE = CreateObject("InternetExplorer.Application")
    objIE.Visible = True
    objIE.Navigate WebAddress
    IEActive
End Function


'Click the Link on web page
```

```
Function WebLink(NavLink)
    Dim i
    Dim TotalLinks
    Dim Linkexist
    Linkexist = False
    TotalLinks = objIE.Document.Links.Length
    For i=0 To TotalLinks-1
        If InStr(LCase(objIE.Document.Links(i).innerHTML),
LCase(NavLink)) > 0 Then
                objIE.Document.Links(i).click
                Linkexist = True
                Exit For
        End If
    Next
    If Linkexist = False Then
        Wscript.Echo "Script could not find this link: " & NavLink
        Wscript.Quit
    Else
        IEActive
    End If
End Function


'Get the current working directory
Function GetPath()
    ' Retrieve the script path
    DIM path
    path = WScript.ScriptFullName ' Script name
    GetPath = Left(path, InstrRev(path, "\"))
End Function


'Create Output txt file
Function CreateOutput(Filename, FileContent)
    Dim fso, OutputFile, MyFile
    OutputFile = GetPath + Filename
    Set fso = CreateObject("Scripting.FileSystemObject")
        If fso.FileExists(OutputFile) = False Then
        Set MyFile = fso.CreateTextFile(OutputFile, True)
        MyFile.Close
        end if
    Set MyFile = fso.OpenTextFile(OutputFile, 8, True)
    MyFile.WriteLine(FileContent)
    MyFile.Close
    Set fso = Nothing
    Set MyFile = Nothing
End Function
```

There are total 4 SQL queries included in the script.

Query I: Get the last date from database which we are going to use it compare the Report Date.

```
SELECT TO_CHAR (MAX (orderdate), 'YYYY-MM-DD')
       AS orderdate
FROM mastdb.customerorder;
```

ORDERDATE
2009-04-30

Query II: Get the daily summery from database.

```
SELECT    storeid AS store_id,
          COUNT (customerid) AS daily_customers,
          SUM (totalpurchase) AS daily_amt
   FROM (SELECT   customerid, storeid,
          SUM (subtotal + order_taxamt) AS totalpurchase
          FROM mastdb.customerorder
          WHERE orderdate = (SELECT MAX (orderdate)
                             FROM mastdb.customerorder)
          -- Make sure customer is counted only once
          GROUP BY storeid, customerid)
GROUP BY storeid
ORDER BY storeid;
```

STORE_ID	DAILY_CUSTOMERS	DAILY_AMT
2001	66	82336.32
2003	33	96543.81
2007	34	22855.38
2012	33	56270.61

Query III: Check how many days in the current month.

```
SELECT COUNT (1)
  FROM (SELECT   mcoo.orderdate
          FROM (SELECT *
                FROM mastdb.customerorder mco,
                (SELECT MAX (orderdate) AS maxorderdate
                  FROM mastdb.customerorder) ldt
                WHERE TO_CHAR (mco.orderdate, 'Year') =
                      TO_CHAR (ldt.maxorderdate, 'Year')
                AND TO_CHAR (mco.orderdate, 'Month') =
```

```
                 TO_CHAR (ldt.maxorderdate, 'Month')) mcoo
GROUP BY mcoo.orderdate);
```

COUNT(1)
30

Query IV: Get the Monthly Summary Amount.

```
SELECT   mcoo.storeid,
           SUM (mcoo.subtotal + mcoo.order_taxamt)
               AS totalpurchase
  FROM (SELECT * FROM mastdb.customerorder mco,
           (SELECT MAX (orderdate) AS maxorderdate
               FROM mastdb.customerorder) ldt
           WHERE TO_CHAR (mco.orderdate, 'Year') =
               TO_CHAR (ldt.maxorderdate, 'Year')
           AND TO_CHAR (mco.orderdate, 'Month') =
               TO_CHAR (ldt.maxorderdate, 'Month')) mcoo
GROUP BY mcoo.storeid
ORDER BY storeid;
```

STOREID	TOTALPURCHASE
2001	2482498.13
2003	2936305
2007	578096.7
2012	1706875.17

Testing Folder:

The automation script is saved into a txt file and name it as *.vbs.

Before test execution:

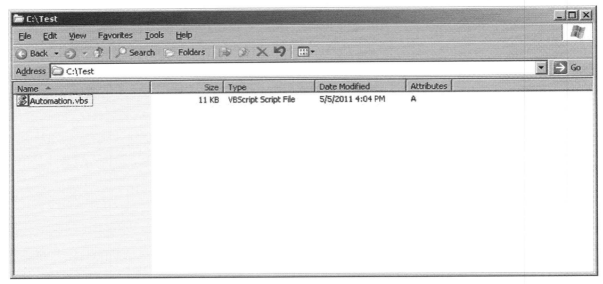

Figure 10.42

After test execution:

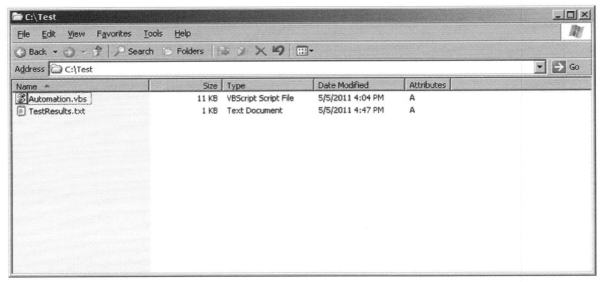

Figure 10.43

Test Result file: All passed

Figure 10.44

or Test Result file: failed on the verification

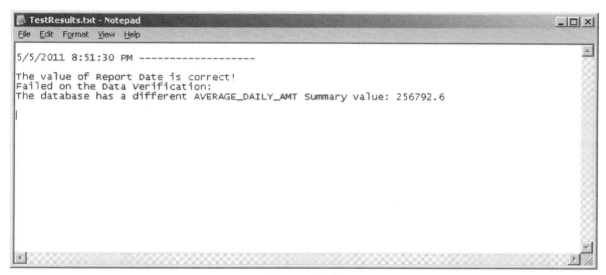

Figure 10.45

A Sample of the DWH Test Plan

A Data Warehouse Test Plan

REVISION HISTORY

The softcopy of the most current version of this Test Plan can be found on Sharepoint.

Version Number	Revision Date MM/DD/YYYY	Summary of Changes	Document Author
0.1		Initial Draft	

Introduction

Project Overview

Below are some of the key benefits for migrating corporation data into DWH:

Integrated	Data is integrated for the whole enterprise to present single version of true.
High Availability	24x7x365 availability with Production fail-over and Disaster Recovery (DR) capability.
Scheduling	Full integration of all corporate data sources based enterprise job scheduler that provides automated and custom scheduling with built-in monitoring.

Monitoring	Full integration with enterprise job monitoring solution that provides enhanced monitoring, reporting & analysis, diagnostics and problem resolution.
Audit	Single point of audit for all data transfers through the use of a centralized service and reporting capabilities.
Improved Supportability	Common process and tools enable simplified support processes and knowledge transfer resulting in increased customer service, quicker problem resolution and less downtime.
Scalability	Simplifies the addition of new feeds or changes to existing feeds.

The overall project objective is to integrate all corporate data from many sources ETL to the single enterprise data warehouse.

This particular test plan will focus on the testing and verification of source data files and through the intermediate stages into final enterprise DWH destination.

QA team will verify the following:

- Verification of source data files and verification of continuous quality monitoring process
- Verification of ETL process
- Full "end-to-end" verification process
- Performance testing
- Regression testing

Test Objectives

The purpose of a combined unit test (UT) system integration testing (SIT) and user acceptance testing (UAT) test plan is to plan the effort required for testing all phases concurrently. The test plan breaks down the business and technical functions into test components, plans staff assignments, determines how the tests will be administered and managed, documents the test facilities and environments to be set-up to run the tests and sets out the schedules for all set-up activities and the running of the tests.

The majority of the technical verification will take place during SIT by the QA Analysts. Where required or in scope, business resources will validate content of the input/output files during the UAT phase.

The objective of the System Integration/User Acceptance testing is to ensure that all DWH components within the scope full tested and to ensure that there are no known defects and that the code is stable and ready for production.

The primary test objectives are to ensure that the:

- The overall strategy for all areas of functional and performance testing.
- Identify test cases for all phases of testing that will be created.
- Define defect tracking procedures.
- Identify testing risks and assumptions.
- Identify required resources for the test effort.
- Provide the testing schedule.
- Confirm that DWH is stable and ready to be deployed to preproduction environment.

The secondary objective is to ensure that the testing is performed in a cost effective and structured manner. This will be accomplished by utilizing automated testing tools—HP Quality Center, HP Quick Test Professional and Excel Macro's where feasible.

To augment the process for successfully implementing data warehouse, there are some additional activities:

- Official Rééquipements (SRS documentation)—validation
- Traceability matrix
- Testing Risk Assessment
- Nonfunctional Testing
- Preproduction Activity Test

During SIT/UAT, the testing will be executed in a test environment. Defects will be documented in the Quality Center Defect Module, indicating Date Opened, Description, Severity, Status, Notes and Action required. The testing will continue until all the SIT/UAT tests are passed in accordance with the SIT/UAT Exit Criteria.

This test approach document describes the appropriate strategies, processes, methodologies used to plan, organize, execute, and manage the testing of the data warehouse and implementation.

Scope

In Scope

The following areas of DWH project are considered in scope of testing in this test plan:

- Data Validation Testing
  - Inbound
  - Outbound
  - Internal Data Movement
- Forms and Assignment Testing
  - Form layout
  - Calculations
  - Security
  - Workflow
- Business Rules Testing
- Performance/Load Testing
- Regression Testing

Out of Scope

The following areas of DWH project are considered out of scope and will not be included in this test plan:

- Dashboard testing & Validation
- Testing External Systems (applications)

Roles and Responsibilities

The following table describes the roles that will be required for successful execution of all tests:

Role	Responsibilities	Resource
Test Lead	Complete test plan and deliver test results; ensure test schedules are met; track all defect reports and their resolutions; deliver test status reports; and serve as an escalation point for all tests.	Name, e-mail, phone
Functional Tester	Design and execute functional test cases; execute all functional tests; write SQL and build pivot table comparisons for data tests; write defect reports and deliver status to test lead.	Name, e-mail, phone

Role	Responsibilities	Resource
Performance Engineer	Define Performance Test Plan; create automated load scripts, execute load tests; monitor test environment; compile performance test results; write bug reports; and deliver status to test lead.	Name, e-mail, phone
ETL Support	An Extract/Transform/Load (ETL) subject matter expert, familiar with all of the business rules surrounding data movement that can answer any questions for the test effort.	Name, e-mail, phone
Database Modeler Support	A Model subject matter expert, familiar with all of the business rules surrounding the form and application design that can answer any questions for the test effort.	Name, e-mail, phone
OLAP Support	A developer familiar with the implementation and business rules implemented in the cube environment	Name, e-mail, phone
Test Environment Support	A support contact for infrastructure issues related to the test environment.	Name, e-mail, phone

Assumptions for Test Execution

The following assumptions are made with respect to the test effort in writing this plan:

- A separate test environment will be supplied and available for the test efforts that meets the following requirements:
    - All test environments must be capable of running all application functionality.
    - The performance test environment must be production-like in terms of load capacity (number of servers, speed of processors, amount of memory, etc.).
    - All test environments must be available and isolated from other use for the tests to be run without interference from ongoing development work.
- Business rules with respect to data movement, security, forms design, and performance are defined, signed off and accessible to the test team.
- The test team is granted the role of QA tracking tool (i.e., HP Quality Center).
- The test team is granted the role of administrator on all test environment servers.

- The test team is granted all required access to source data systems, and outbound data streams.
- Assignments must be granted to the test team members for all relevant forms.
- Development is complete and has been unit tested for an area of functionality prior to the beginning of the test execution period for that area of functionality.

Risks and Risk mitigation

Besides the regular software testing risks like completion of requirements documentation, and development prior to test execution, availability of the test environment, and recourses. the following are specific risk when testing DWH:

Some risks and risk mitigation in testing data warehouse are listed below				
Risk ID	Risk	Description and cause of risk	Probability [H - High M - Mid, L - Low]	Risk mitigation
R01	User heterogeneity	Many uses with different requirements	H	CEO champion
R02	Growth and scalability	There will be more users and more source files	H	Verify scalability of design
R03	System evolution	New requirements new "appetites"	H	Anticipate an verify extensibility
R04	System integration from many sources	Many version of truth need to be converted into a single version of truth	H	Extensive verification of dada mapping document
R05	Availability of experienced practitioners	DWH has yet to mature as a discipline, adding to it the inherent complexity of DWH	H	Institute training and education

	Some risks and risk mitigation in testing data warehouse are listed below			
R06	Pressure for delivery of effective DWH solutions	Ever increasing competition, more sophisticated and better informed consumers, changing markets	H	Rigorous and comprehensive assessment leading to successful incremental delivery, targeting first delivery within 6 weeks. [6 weeks is the upper tolerance limit of the business users]
R07	Stakeholders' expectations in the DWH?	DWH represents single version of truth. Each stakeholder's version must be included	H	List of stakeholders and their expectation
R08	Scope of delivery	The iterations have been previously agreed upon between Development, Business and QA team.	H	Establish clarity and consensus on the scope of each iteration, exact deliverables, and expected outcome of the assessment
R09	Tight Alliance between QA Business team	Position DWH as a strategic business asset with the implication of the business knowledge management. Has an active and committed business sponsor been identified?	H	Identify business owner as the person responsible for delivering DWH solution. With a focus on delivering the business value.
R10	Technical Architecture supports Information Architecture	Business information can be easily extracted from the underlying technical architecture	H	Early paper test of implementation of the business rules using "what if" and exercising each business rule.
R11	Understanding how DWH relates to other organization initiatives	Both those that feed DWH and those than extract the information from it have to be aware of the new roles.	H	Obtain the commitment for data sourcing departments for the continuous data quality improvements efforts

Some risks and risk mitigation in testing data warehouse are listed below				
R12	Preparation to transition from warehouse "builders" to warehouse "operators"	In process of building DWH the future need for support of both operation and enhancements must be anticipated and preparation should the be made in the period	H	Plan for support of DWH operation.

Test Methodology

Data Validation Testing

There are three categories of data validation, all of which will require their own set of test cases.

Inbound Data Validation Testing

There are separate ETL feeds for each source system that provides data to the Planning Server and the Monitoring Server. Each source system requires a test case to validate that the data in the Planning Server and the Monitoring Server ties to the source data and conforms to any rules that exist to transform the source data prior to loading it to the Planning Server and the Monitoring Server.

The following table lists the test cases that will be developed by the test team to validate all inbound data:

Data Source	Test Case Name	Test Case Description	Test Case Methodology
Source System 1	Inbound Data 1	Example: Validates all Budget dollar values inbound from source system 1	SQL
Source System 2	Inbound Data 2	Example: Validates all Headcount data inbound from source system 2	Excel Pivot comparisons

Outbound Data Validation Testing—Planning

There are separate ETL feeds for each outbound data stream, and each outbound feed requires a test case to validate that the data in the Planning Server ties to the outbound data feed and conforms to any rules that exist to transform the source data prior to sending the data.

The following table lists the test cases that will be developed by the test team to validate all outbound data:

Outbound Feed	Test Case Name	Test Case Description	Test Case Methodology
Outbound feed 1	Outbound Data 1	Example: Validates all Budget dollar values sent in outbound feed 1	SQL
Outbound Feed 2	Outbound Data 2	Example: Validates all Headcount data sent in outbound feed 2	Excel Pivot comparisons

Internal Data Movement Validation Testing—Planning

There are separate ETL feeds that move and transform data between models. Each movement requires a test case to validate that the data loaded to the Planning Server target ties to the source data and conforms to any rules that exist to transform the source data prior to loading it to target.

The following table lists the test cases that will be developed by the test team to validate all internal data movements:

Outbound Feed	Test Case Name	Test Case Description	Test Case Methodology
Internal Data Movement 1	Internal Data 1	Example: Validates that all data in Model 1 ties to the data in Model 2, with the geography hierarchy in Model 2 rolled up 2 levels from the hierarchy in Model 1	SQL
Internal Data Movement 2	Internal Data 2	Example: Validates that all data moved from Model 2 to Model 3, with Revenue from Model 2 allocated based on the percentages provided in the allocation data feed as it is moved into Model 3.	Excel Pivot comparisons

Internal Data Movement 3	Internal Data 3	Example: Validate that the month-end close process successfully archives the forecast data, and sets the current period one month forward.	SQL

Rules Testing—Planning

Each rule must be tested to verify the data and calculations effected by the rule tie to requirements. The following table lists the test cases that will be developed by the test team to validate all Rules

Rule	Type	Test Case Name	Test Case Description	Test Case Methodology
Rule 1	Definition	[Rule 1]	Example: Verify that the Calculated Member "Variance_ActualsVsBudget" matches the formula Actuals—Budget	Review values in the form
Rule 2	Assignment	[Rule 2]	Example: Verify that the Cost per Head member correctly calculates based on the formula Operating Expenses / Total Headcount	Review values in the form

Performance Testing—Planning

The requirements state that the application must support an active concurrent user base of "Number of users that will be active on the system concurrently during the peak hour" users.

Test case to simulate the user scenario and number of used is identified with tool that would be used to execute user scenarios (example HP Load Runner).

Performance Test Monitoring

The requirements stated that the application must also support a data refresh rate meeting certain criteria from source to cubes.

Additionally, the requirements state that the application must support an active concurrent user base of "number of users that will be active on the system concurrently during the peak hour" users.

Providing the level of load on the servers, the response times cannot exceed what is represented in the following table:

Transaction	Required Response Time
ETL Process Run [source group 1]	<4 hours
Cube Update [Cube 1]	<4 hours
Open [Report 1]	5 seconds
Open [Pivot 1] if pivot are used	15 seconds
Refresh data in [Report 1]	<5 seconds
Refresh data in [Pivot 1]	<15 seconds

ETL Process Testing

ETL Process Testing has the goal of ensuring the data being loaded from external data sources into the target system meets the overall data refresh rate specified in the requirements. The objective of this test is to test the ETL scenario with the expected data volumes (including growth) over the next twelve months of both source and target systems.

Regression Testing

All the tests above are intended to ensure that functionality is complete and is working according to the requirements. Once the initial testing is complete, though, bugs are discovered and fixes are implemented, and change requests are completed. As that happens, the modified code must be tested and retested. However, it is also necessary to test that the modifications made did not in any way break the other previously tested code. So a set of critical path regression tests must be executed after every change to application functionality before the application can be released to UAT and/or production.

The following table lists the critical path regression test cases that will be executed after any modifications to the code.

Test Planning

Test Case ID	Description
Inbound Data 1	Example: Validates all Budget dollar values inbound from source system 1
Inbound Data 2	Example: Validates all Headcount data inbound from source system 2
Outbound Data 1	Example: Validates all Budget dollar values sent in outbound feed 1
Outbound Data 2	Example: Validates all Headcount data sent in outbound feed 2
Internal Data 1	Example: Validates that all data in Model 1 ties to the data in Model 2, with the geography hierarchy in Model 2 rolled up 2 levels from the hierarchy in Model 1
Internal Data 2	Example: Validates that all data moved from Model 2 to Model 3, with Revenue from Model 2 allocated based on the percentages provided in the allocation data feed as it is moved into Model 3.
Internal Data 3	Example: Validate that the month-end close process successfully archives the forecast data, and sets the current period one month forward.
Rule 1	Example: Verify that the Calculated Member "Variance_ActualsVsBudget" matches the formula Actuals—Budget
Rule 2	Example: Verify that the Cost per Head member correctly calculates based on the formula Operating Expenses / Total Headcount
Performance / Stress Test	Gradually increase concurrent user load to determine the breaking points of the test environment with the application.

Suspension/Resumption Criteria

Testing will be suspended and resumed according to the following conditions:

Suspension Criteria	Resumption Criteria
All test cases have been run as close to completion as possible and are awaiting bug fixes for retest	Testing will resume once the development team has resolved the bug and assigned it back to the test team

Testing is not possible due to a failure of a system or deliverable spelled out as a dependency in the test plan or cases	Testing will resume once the appropriate support team closes the issue raised by the test team
Critical scope changes occur which require a revision of the test plan and/or cases which will require approval from test stakeholders	Testing will resume once all scope changes have been built into the test plan and cases, and the modifications to the test plan and cases have been reviewed and approved by all test stakeholders

Test Completeness Criteria

Testing will be considered complete when the following conditions are met:

- All tests have been executed.
- All priority 1 and 2 bugs have been resolved.
- All test stakeholders agree that based on the test results the requirements have been acceptably met.

Test Deliverables

The following are deliverables from the testing effort:

Test Plan Document

The test plan document will be written by the Test Lead. The purpose of the Test Plan document is to:

- Specify the approach that testing will use to test the product, and the deliverables.
- Break the product down into distinct areas and identify features of the product that are to be tested.
- List all test cases that will be developed to test the application.
- Specify the procedures to be used for testing sign-off.
- Indicate the tools used to test the product.
- Indicate the contact persons responsible for various areas of the project.
- Identify risks and contingency plans that may impact the testing of the application.
- Specify bug management procedures for the project.

Test Schedule Document

The tests schedule will be developed by the Test Lead, and will be based on information from the Project Manager.

Test Cases Documentation

Test cases will be designed and documented for each test case specified in the Test Plan document. They will be designed and documented by the testers for the respective cases.

Functional and Performance Test Reports

The results of the Functional and Performance Tests will be written up in a report and delivered to all test stakeholders. The report will detail the passing and failing test cases, and all bugs open and resolved from the testing effort.

Test Status Reports

A weekly status report of the test efforts will be distributed to all test stakeholders, detailing the number of test cases that have been executed, the pass/fail ratios, all open bugs, and all bugs resolved since the last status report.

Defect Tracking

Defect Tracking System

Different client have different defect tracking systems. HP Quality Center is assumed here; if there is none, the test lead will maintain the list of defects and their status in Excel, and distribute the reports as required.

Defect tracking system should implement process as presented on the flow diagram below:

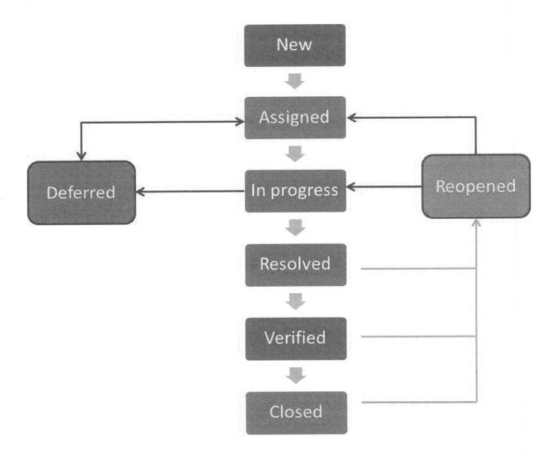

Defects Triage

The test lead and development lead, and any other personnel deemed necessary on a meeting by meeting basis, will review open bugs in a daily triage session during test execution, to determine status and assign priority and severity for the bugs. Planning the triage meetings will be the responsibility of the test lead.

The test lead will provide required documentation and reports on bugs for all attendees. Development will then assign the bugs to the appropriate person for fixing and report the resolution of each bug. The test lead will be responsible for tracking and reporting on the status of all bug resolutions.

Defect Severity and Priority Definitions

Bug Severity and Priority fields are both very important for categorizing bugs and prioritizing if and when the bugs will be fixed. The bug severity and priority levels will be defined as outlined in the following tables below. Testing will assign a severity level to all bugs. The test lead will be responsible to see that a correct severity level is assigned to each bug.

Severity List

Severity ID	Severity	Description
1	Critical	The application crashes or the bug causes nonrecoverable conditions. System crashes, or database or file corruption, or potential data loss, program hangs requiring reboot are all examples of a Sev. 1 bug.
2	High	Major system component unusable due to failure or incorrect functionality. Sev. 2 bugs cause serious problems such as a lack of functionality, or insufficient or unclear error messages that can have a major impact to the user, prevents other areas of the app from being tested, etc. Sev. 2 bugs can have a work around, but the work around is inconvenient or difficult.
3	Medium	Incorrect functionality of component or process. There is a simple work around for the bug if it is Sev. 3.
4	Minor	Documentation errors or signed off severity 3 bugs.

Priority List

Priority ID	Priority	Description
5	Must Fix	This bug must be fixed immediately; the product cannot ship with this bug.
4	Should Fix	These are important problems that should be fixed as soon as possible. It would be an embarrassment to the company if this bug shipped.
3	Fix When Time Allows	The problem should be fixed within the time available. If the bug does not delay shipping date, then fix it.
2	Low Priority	It is not important (at this time) that these bugs be addressed. Fix these bugs after all other bugs have been fixed.
1	Trivial	Enhancements/Good to have features incorporated—just are out of the current scope.

Testing Tools

The following tools will be used in the test execution process:

- HP Quality Center or equivalent test management tool
- SQL Server Management Studio
- HP Quick Test Pro or equivalent
- Excel

Test Plan Approver List

The following list represents the stakeholders for this testing effort. The people identified in this list must approve the test plan, will receive all test status reports throughout the testing effort, are required to approve any changes to the test plan or scope, and will be included on the distribution list for the test results reports.

Name	Role	E-Mail Address	Phone

Documentation

The following documentation was utilized to create this test plan.

Document Name	Document Location

Distribution

The softcopy of the most current version of this test plan has been distributed to the following individuals.

Name	Title	Department	Review ✓	Copy ✓

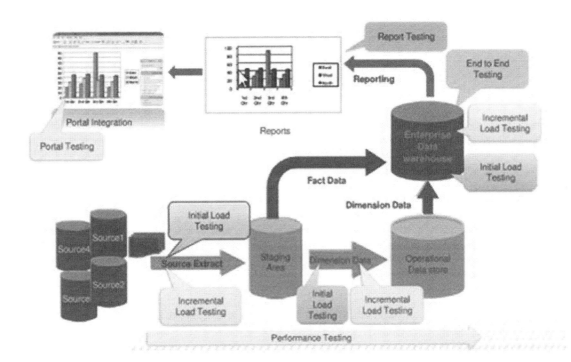

Dr. Deming's 14 Points:

1. *Constancy of purpose*
   Create constancy of purpose for continual improvement of products and service to society, allocating resources to provide for long range needs rather than only short term profitability, with a plan to become competitive, to stay in business, and to provide jobs.

2. *The new philosophy*
   Adopt the new philosophy. We are in a new economic age, created in Japan. We can no longer live with commonly accepted levels of delays, mistakes, defective materials, and defective workmanship. Transformation of Western management style is necessary to halt the continued decline of business and industry.

3. *Cease dependence on mass inspection*
   Eliminate the need for mass inspection as the way of life to achieve quality by building quality into the product in the first place. Require statistical evidence of built in quality in both manufacturing and purchasing functions.

4. *End lowest tender contracts*
   End the practice of awarding business solely on the basis of price tag. Instead require meaningful measures of quality along with price. Reduce the number of suppliers for the same item by eliminating those that do not qualify with statistical and other evidence of quality. The aim is to minimize total cost, not merely initial cost, by minimizing variation. This may be achieved by moving toward a single supplier for any one item, on a long term relationship of loyalty and trust. Purchasing managers have a new job, and must learn it.

5. *Improve every process*
   Improve constantly and forever every process for planning, production, and service. Search continually for problems in order to improve every activity in the company, to improve quality and productivity, and thus to constantly decrease costs. Institute innovation and constant improvement of product, service, and process. It is management's job to work continually on the system (design, incoming materials, maintenance, improvement of machines, supervision, training, retraining).

6. *Institute training on the job*
   Institute modern methods of training on the job for all, including management, to make better use of every employee. New skills are required to keep up with changes in materials, methods, product and service design, machinery, techniques, and service.

7. *Institute leadership*
    Adopt and institute leadership aimed at helping people do a better job. The responsibility of managers and supervisors must be changed from sheer numbers to quality. Improvement of quality will automatically improve productivity. Management must ensure that immediate action is taken on reports of inherited defects, maintenance requirements, poor tools, fuzzy operational definitions, and all conditions detrimental to quality.

8. *Drive out fear*
    Encourage effective two way communication and other means to drive out fear throughout the organization so that everybody may work effectively and more productively for the company.

9. *Break down barriers*
    Break down barriers between departments and staff areas. People in different areas, such as Leasing, Maintenance, Administration, must work in teams to tackle problems that may be encountered with products or service.

10. *Eliminate exhortations*
    Eliminate the use of slogans, posters and exhortations for the work force, demanding zero defects and new levels of productivity, without providing methods. Such exhortations only create adversarial relationships; the bulk of the causes of low quality and low productivity belong to the system, and thus lie beyond the power of the work force.

11. *Eliminate arbitrary numerical targets*
    Eliminate work standards that prescribe quotas for the work force and numerical goals for people in management. Substitute aids and helpful leadership in order to achieve continual improvement of quality and productivity.

12. *Permit pride of workmanship*
    Remove the barriers that rob hourly workers, and people in management, of their right to pride of workmanship. This implies, among other things, abolition of the annual merit rating (appraisal of performance) and of management by objective. Again, the responsibility of managers, supervisors, foremen must be changed from sheer numbers to quality.

13. *Encourage education*
    Institute a vigorous program of education, and encourage self improvement for everyone. What an organization needs is not just good people; it needs people that are improving with education. Advances in competitive position will have their roots in knowledge.

14. *Top management commitment and action*
    Clearly define top management's permanent commitment to ever-improving quality and productivity, and their obligation to implement all of these principles. Indeed, it is not enough that top management commit themselves for life to quality and productivity. They must know what it is that they are committed to—that is, what they must do. Create a structure in top management that will push every day on the preceding thirteen points, and take action in order to accomplish the transformation. Support is not enough; action is required.

The above query returns all rows in either table that do not completely match all columns in the other. In addition, it returns all rows in either table that do not exist in the other table. It handles nulls as well, since GROUP BY normally consolidates NULL values together in the same group. If both tables match completely, no rows are returned at all.

Endnotes

1. Ralph Kimball and Margy Ross, *The Data Warehouse Toolkit: The Complete Guide to Dimensional Modeling* (Wiley & Sons).

2. Dean Leffingwell and Don Widrig, *Managing Software Requirements A Use Case Approach* (2003).

3. Donald Gause and Gerald Weinberg, *Exploring Requirements: Quality before Design* (New York: Dorset House Publishing, 1989).

4. W.H. Inmon, *Building the Data Warehouse* (New York: John Wiley and Sons, September 1998).

5. Alan Cooper, *The Inmates Are Running the Asylum.*

6. Edward Tufte, *The Visual Display of Quantitative Information,* 2nd edition.

7. Stephen Few, *Show Me the Numbers: Designing Tables and Graphs to Enlighten.*

8. Stephen Few, *Information Dashboard Design: The Effective Visual Communication of Data.*

9. Dr. William S. Cleveland, *The Elements of Graphing Data.*

10. Steven D. Levitt and Stephen J. Dubner, *Freakonomics.*

11. Donald A. Norman, *Things That Make Us Smart: Defending Human Attributes in the Age of the Machine.*

12. Dr. John Maeda, *The Laws of Simplicity (Simplicity: Design, Technology, Business, Life).*

13. Marty Neumeier, *The Brand Gap: Expanded Edition.*

14. Daniel H. Pink, *A Whole New Mind: Why Right-Brainers Will Rule the Future.*

15. Dr. Jacques Bertin, *Semiology of Graphics: Diagrams, Networks, Maps.*

16. Dr. Colin Ware, *Information Visualization, Second Edition: Perception for Design* (Interactive Technologies).

17. Chip Heath and Dan Heath, *Switch: How to Change Things When Change Is Hard.*

18. Chip Heath, *Made to Stick: Why Some Ideas Survive and Others Die.*

19. Mackinlay Card and Shneiderman, *Readings in Information Visualization.*

20. Dr. Ian Ayres, *Super Crunchers.*

21. Toby Segaran and Jeff Hammerbacher, *Beautiful Data: The Stories Behind Elegant Data Solutions.*

22. David Levy, *Tools of Critical Thinking.*

23. Dr. Betty Edwards, *Drawing on the Right Side of the Brain.*

24. Dr. Betty Edwards, *Drawing on the Artist Within.*

25. Tony Schwartz, *What Really Matters: Searching for Wisdom in America.*

26. Michael Michalko, *Thinker Toys,* 2nd edition.

27. Kelsey Ruger, *The Owner's Manual for the Brain*

28. Dr. Boris Beizer, *Software Testing Techniques.*

29. Chic Thompson, *What a Great Idea.*

30. Peter Drucker, *The Age of Discontinuity.*

31. Plato, *The Republic.*

32. David Hand, Heikki Mannila and Padhraic Smyth, *Principles of Data Mining.*

33. Dr. Andrew Abela, *The Presentation: A Story about Communicating Successfully with Very Few Slides.*

34. Dr. Viktor Frankl, *Man's Search for Meaning.*

35. Dr. Howard Wainer, *Visual Revelations: Graphical Tales of Fate and Deception from Napoleon Bonaparte to Ross Perot.*

36. Dr. Howard Wainer, *Graphic Discovery: A Trout in the Milk and Other Visual Adventures.*

37. Dr. Howard Wainer, *Picturing the Uncertain World: How to Understand, Communicate and Control Uncertainty through Graphical Display.*

38. Dr. L. Wilkinson, *The Grammar of Graphics.*

39. Dr. Boris Beizer, *Software System Testing and Quality Assurance.*

40. Dr. Mihaly Csikszentmihalyi, *Flow: The Psychology of Optimal Experience.*

41. Humphrey Watts, *Managing the Software Process.*

42. Steve Sarsfield, *The Data Governance Imperative.*

43. Thomas Redman, *Data Driven: Profiting from Your Most Important Business Asset.*

44. Jack E. Olson, *Data Quality, the Accuracy Dimension* (Morgan Kaufmann, 2003).

45. Arkady Maydanchik, *Data Quality Assessment* (Technics Publications, 2007).

46. Leander Kahney, *Inside Steve's Brain.*

47. Daniel Levitin, *This Is Your Brain on Music.*

48. John Medina, *Brain Rules: 12 Principles for Surviving and Thriving at Work, Home, and School.*

49. Raj Kamal, *Microsoft Corporation: Resurrecting the Prodigal Son—Data Quality.*

50. Jaiteg Singh, *Statically Analyzing the Impact of Automated ETL Testing on Data Quality of a Data Warehouse.*

51. Vincent Rainardi, *Building a Data Warehouse with Examples on SQL Server.*

52. Eric Ceres, *RTTS Corp.: Ensuring Pharmaceutical Safety Data Integrity in a Data Warehouse.*

53. Infosys, *Clearware: A Data Warehouse Testing Solution.*

30502764R00170

Made in the USA
Lexington, KY
06 March 2014